T0093634

WHEN THE LAND TURNED GREEN

The Maine Discovery of the First Land Plants

Dean Bennett and Sheila Bennett

Down East Books
CAMDEN, MAINE

Down East Books

Published by Down East Books
An imprint of Globe Pequot, the trade division of
The Rowman & Littlefield Publishing Group, Inc.
4501 Forbes Blvd., Ste. 200
Lanham, MD 20706
www.rowman.com

Distributed by NATIONAL BOOK NETWORK

British Library Cataloguing in Publication Information available

Library of Congress Cataloging-in-Publication Data on file

ISBN 978-1-68475-032-0 (hardcover)
ISBN 978-1-68475-033-7 (e-book)

To the paleontologists and geologists
who revealed the Trout Valley Formation's secrets
of how plant-life evolved on Earth

CONTENTS

PREFACE

DEEP IN THE WILDERNESS OF NORTHERN MAINE IN THE MID-1950S, A Harvard PhD student is wading down a mountain stream into a remote valley. He is wading because he is looking for rocks the stream has revealed in its bed and along its edges, the avenue it has cut through the thick tangled forest that covers most of the land here. He is taking his first steps to map the geology of 300 square miles of Baxter State Park, a daunting task that will take him six years to complete. But he must see the rocks to do this, for they can tell him the geological secrets of this place. He finds his way around a waterfall and beyond he discovers a series of unusually-shaped rock outcrops. He learns later that they are part of an unknown geologic formation, hundreds of millions of years old, still mystifying today because of its relative lack of change despite nearby volcanic activity and massive land movement. Wading on, he has another surprise. In a thin layer of black shale beside the stream, he finds a small fossil of a plant. Yet many others have visited these woods and streams before him with no public comment of such a finding.

How much the young scientist was aware of the full significance of his discovery of the formation and its fossils that day in 1955 we do not know, but a fascinating story of science was about to unfold. It is a story that needs to be told, and never more so than today when our lives and future existence depend so much on the work of scientists as well as a recognition of the value of their work to society. Not since Darwin's time has science been faced with so many challenges from society; nor has it offered so much promise for the future. As science educators, we learned the details of this story through our years of teaching and writing and exploring this area with scientists. We believe that it will capture the interest of our readers and further their understanding of our planet as much as it has ours.

In our book, we tell how scientists revealed the details of a major event in the history of our planet. This event was the transition of plants to land. It is an occurrence that had, and continues to have, a critical influence on the Earth's life-supporting processes, including our climate. We do this by tracing the story of the discovery of one of the world's ancient land plants. As far as we know, the fossilized remains of the plant known as *Pertica* (genus) *quadrifaria* (species) are found only in that Devonian-age geological formation residing in what is still a wild, remote, and heavily-forested place in northern Maine. The place is now protected within more than 200,000 acres deeded to the State of Maine by its donor, former Maine Governor Percival P. Baxter, whose name is forever a part of the land area known as Baxter State Park.

The fossil plant was alive about 398 million years ago during the Early Devonian Period of geologic time. It was one of the first land plants and is thought to have been the tallest plant on the landscape in its time. Today, the site of the fossil's discovery lies in the shadow of an Eastern White Pine, *Pinus strobus,* also the Maine State Tree. The pine now takes the ancient plant's place as the tallest plant on the land in the eastern part of the United States.

In this book we explore the hypothesis that these two tall plants are related. We also ask: What can these two plants, one ancient, and one modern, tell us about the past and perhaps hint at the future? And how is it that they both met here after the fossil had traveled thousands of miles from the southern hemisphere where it once existed? And why is it that both find themselves in a region divided by two very different ideas about our relationship and responsibility to the land—to keep it forever wild or to manage it for its sustainable use?

The fossil plant, *Pertica quadrifaria*, is now the official Maine State Fossil, and we tell the story of how this came to be. It is a plant devoid of leaves but suggesting an evolutionary promise of trees with leaves. It is a plant whose complex form is now strikingly displayed by a three-dimensional drawing, all because of the painstaking work of the scientist who discovered it.

Beyond the symbolic recognition and societal values of the fossil and the tree, there is another story. It is about the place where they are found—within or anchored in a distinct geologic formation, exposed as a belt a few miles long, relatively narrow, and hundreds of feet deep. Named the Trout Valley Formation, it is called an enigma by geologists because it was, relatively speaking, strangely unchanged by the immense geological forces of land movement and volcanic eruption that once occurred in its surroundings. Today, this massive rock formation is covered by a deep forest, except where flowing brooks and streams, and limited human activities, have exposed its forested surface.

The story is also about those who revealed the history of the fossil and the pine and the place where the two reside. They are the scientists: the geologists, biologists, and other scientists, but especially the paleontologists. It is this group who integrate botany, geology, ecology, climatology, and other sciences to understand the history of life on Earth. This story could not be told without them. They worked, and continue to do so, to unravel the fascinating history of Maine preserved here. They follow their curiosity, the motivation that drives science, leading them to investigate the ancient past and try to predict the future. With modern methods and an ever-changing technology, they search, discover, study, and communicate their findings to help us better understand our place and role on Earth.

Our book has been a long time in the making. It was in 1987 that one of us, Dean, visited one of our country's highly respected and well-liked paleontologists, Professor Henry Andrews, at his retirement home, a family farm in New Hampshire. Dr. Andrews was one of the discoverers of *Pertica quadrifaria*. The purpose of the visit was to interview him for a potential children's book. After lunch and the interview, Professor Andrews invited Dean out into the barn of his homestead. Inside, he walked over to a well-used workbench, brushed the dust off a large slab of rock, and showed it to Dean. It was a magnificent specimen of the Maine State Fossil, one of museum quality. Then he gave it to Dean, saying: "You can use this better than I can."

Since that time, though the children's book has yet to be published, we, as college teachers, have both brought the story of the fossil to thousands of science teachers and general college students. We have relayed that story in courses we have taught in classrooms, in laboratories, and at a distance through statewide instructional television. We have brought the story to a wider audience online. We also have lent the fossil to other educators to further its educational use. Now, a book once envisioned for children has morphed into this book for the general public. The effort is the result of the gift of a fossil more than thirty years ago.

For those of you who might wish to visit Baxter State Park where the Maine State Fossil and the Trout Valley Formation are located, we have an important note. This is not a guidebook for locating fossils in the Park. There are no maps and directions to fossil sites. Collecting fossils, rocks, and minerals and the use of geologic hammers are strictly forbidden in the Park. Research studies within Baxter State Park may be approved but require a special use permit. Those interested in conducting research should consult Park regulations and personnel. In visits to the Park, safety guidelines provided by the Park should be heeded for it is a remote area and a large wilderness.

ACKNOWLEDGMENTS

WE ARE NOT PALEONTOLOGISTS NOR ARE WE GEOLOGISTS; WE ARE EDUcators who have an in-depth background in the natural sciences. However, if it were not for scientists, we would not have had the courage, background, and support that we drew on to write this book. It was because of their generous spirit of sharing that this book exists—a spirit that appears in this story many times.

First, we thank Andrew E. Kasper, Jr., the only living member of the scientists who were involved in the discovery and study of the Maine State Fossil, *Pertica quadrifaria*. We could not have told this story if it were not for the day he devoted to recounting details of his discovery to one of us (Dean) in demonstrations in his laboratory at Rutgers University in New Jersey in the mid-1980s, for the time he spent several years ago reviewing his knowledge of the fossil in the dining room of Warrena Forbes, the widow of Bill Forbes, in Washburn, Maine, and more recently for the many hours he gave in correcting and adding substantially to our manuscript and giving us photographs and other illustrations to use.

We also owe a debt of gratitude to the late Henry N. Andrews, Jr., who, in the 1960s, left his laboratory and classroom at the University of Connecticut, and led his team of Devonian fossil hunters to the Trout Valley area in Baxter State Park where the Maine State Fossil and other Devonian plant fossils were found. When one of us (Dean) interviewed him in the mid-1980s, he gave him a magnificent, museum-quality specimen he had found.

And to another, the late William H. (Bill) Forbes, also a member of Andrews' team involved in the discovery of the Maine State Fossil and other fossils, who became a professor at the University of Maine at Presque Isle, we extend another special thank you. We also express our appreciation and thanks to his widow, Warrena Forbes, who so warmly and generously helped us understand him and his role as well as hers in the story we tell. She gave us helpful details of his background, of her life married to a paleontologist, many photos of his life and accomplishments, and time and thought to relevant parts of the manuscript.

To Robert G. Marvinney, State Geologist with the Maine Geological Survey, we extend a special thank you for supporting us since the beginning of this writing project and for critically reading the entire manuscript and giving us direction on it. He also provided us key contacts and helped in our search for others who could assist us. His staff also provided much-needed help in providing informational resources.

Another scientist, Colby College geologist and paleontologist Robert A. Gastaldo, was especially helpful with our chapter about the second round of studies of the Trout Valley Formation and its Devonian fossils. For this we are most appreciative and thank him.

We also thank paleobotanist Patricia G. Gensel, professor at the University of North Carolina at Chapel Hill. She began her long career in the 1960s serving on Henry Andrew's fossil-hunting team. She gave us valuable materials on fossils of the Devonian Period and helped to provide two significant photographs for our book: one of her looking for fossils in an outcrop, which was given to us to use by its photographer, James E. Mickle, and another of her with Robert Gastaldo at a fossil site beside Trout Brook.

Another major scientist in our story is the late Douglas Whiting Rankin of the United States Geological Survey, whom we, unfortunately, never personally met but were privileged to be in the group he led on a field trip down South Branch Ponds Brook in the mid-1990s. Although he never knew it, the trip was an important inspiration to us to do this book. We thank the Harvard University Pusey Library and reference staff of the Harvard University archives for their help in obtaining information and scans of Douglas Rankin's original study, photographs, and maps. Their help also led to permissions from Andrea Rankin and Katharine Rankin of Rankin's family to publish relevant materials in our book. To this end, we also thank Robert G. Marvinney and Robert D. Tucker for their help in our attempts to contact the Rankin family.

We also thank Walter A. Anderson, former Maine State Geologist, who helped us with the chapter on the designation of a Devonian fossil plant, *Pertica quadrifaria*, as the Maine State Fossil. He had a major role in the fossil's selection and the subsequent legislation.

The late Robert S. Newman, a research paleontologist for the Smithsonian Institution, was most helpful to one of us (Sheila) to gain field knowledge and understanding of fossils in northern Maine. This occurred in the 1990s when she had been awarded a Libra professorship to study ancient environments of Maine through fossil evidence. Dr. Newman spent several days in the field with her showing her fossil sites adjacent to Rankin's study area in Baxter State Park.

We also express our appreciation and thanks to Jensen Bissel, former director of Baxter State Park, who gave us his support at the beginning of our project before his retirement. We also thank Eben Sypitkowski, another former Director of Baxter State Park and Park Naturalist Marc Edwards for their support of the book project, including their invitation for us to give a PowerPoint presentation to the Park's Research Committee.

For special assistance in obtaining information and photographs, we thank Rebecca Chasse of the University of New Hampshire Archives for helping us obtain use of a 1953 photo of Douglas Rankin with members of an Appalachian Mountain Club trail crew. We also thank Deborah Shapiro of the Smithsonian Institution Archives for assisting us in obtaining correspondence relevant to William H. Forbes' initial appointment to teach and conduct research at the University of Maine at Presque Isle. For these materials we give the following citation: Smithsonian Institution Archives, Accession 15-182, Box 2, Folder 49: Forbes, William H., 1961-1970. We thank geologist Yvette Kuiper of the Colorado School of Mines for use of her photo of Rankin on Mount Evans in Colorado. We thank Professor Emeritus Duncan Heron, geologist at the School of the Environment at Duke University, for use of his photo of Rankin leading a field trip for the Carolina Geological Society. We thank Sarah Keen of Colgate University for helping us obtain permission to publish Rankin's photo in Colgate University's 1953 yearbook *Salmagundi.* We are indebted to Sofia Yalouris, Image Services Coordinator, Maine Historical Society, for her help in obtaining permission to publish a photo of Olof Nylander from the Collections of the Nylander Museum, courtesy of VintageMaineImages.com. We appreciate the assistance of Melissa Bilyeu, Operations Director, of Witherspoon Media Group for obtaining use of a photo of Erling Dorf. We thank Melanie Mohney of the Maine State Library for her help in obtaining photographs of Trout Brook Farm and Governor Joseph Brennan signing the Maine State Fossil legislation. And we appreciate the efforts of Heather Moran, of the Maine State Archives, in our search to find those photographs. Our thanks go, also, to the staff of the Katz Library at the University of Maine at Augusta for searching out scientific papers and books for our research.

We also thank Lindsay and Michael Downing, proprietors of Mt. Chase Lodge; the staff of Shin Pond Camps; and former owners of Camp Wapiti, the Ramellis, for their hospitality during our stays in the area to conduct research for our book

And to Scott Herrick, who kept our computers going during the project, we will be forever grateful for his interest and expertise.

We also acknowledge the influence of the following advisors in our graduate programs. Sheila expresses her thanks to Professor August J. DeSiervo of the University of Maine, and Dean thanks the late Professor Robert Miller of the University of Southern Maine and the late Professor William B. Stapp of the University of Michigan

It goes without saying that we appreciate the support our children have given us, and, especially, we thank Cheryl Bennett Martin and Charles O. Martin for spotting us on our expeditions wading down Trout Brook and South Branch Ponds Brook in Baxter State Park as we attempted to follow in the footsteps of the paleontologists we have written about. And we thank them for permission to include their photos in the book and Rick Bennett to use a photo of him.

PROLOGUE

An ocean wave forms, lengthens out around the cove and then cascades to shore. Its crest of foam washes over a colony of plants growing on the bank of an emerging stream near an eroding volcano on the coastline of a subtropical ocean we know today as the Atlantic. Meanwhile, a warm breeze caresses the plant's branches carrying moisture and carbon dioxide to the cells of stems responsible for growth. The plant, given the name *Pertica quadrifaria*, is the tallest plant on the landscape. It grows with others of its kind in a patch of green on the bank and across the wetlands. There are no leaves on the plant's branches, which spiral up a tall, rod-like stem. The assemblage catches the warm, humid air and sunlight and produces the necessary ingredients for life through a process we call photosynthesis.

Hundreds of millions of years ago before *Pertica*, the continental crust was barren awaiting colonization by algae and other nonvascular plants. Photosynthetic algae in the group known as steptophytic algae had evolved the necessary processes and anatomy for them to move from the water to the land. Their move to land required periodic access to water for fertilization of reproductive cells producing the next generation. The land-based algae enjoyed the direct sunlight and direct access to carbon dioxide in the atmosphere. These first plants colonized the barren earth, and their success led to offspring with the necessary features for survival and reproduction in this new arena of habitation. At first the early land plants hugged the ground, but over time began to stretch upward for the energetic wavelengths of sunlight. Their height shaded the low growing plants which in turn also began to reach for the sunlight.

Ultimately, taller plants evolved; their height, possibly up to a few feet, is witnessed in the plant *Pertica* growing in that patch on the bank of a stream channel in a coastal area called a delta today. The stream drains a steep volcanic slope. And, as so often happens, a rainstorm sweeps over the exposed land, washing mud and debris down the slope onto the delta, burying the patch of plants. It would be hundreds of millions of years later, in middle of the twentieth century, before a species of animal, *Homo sapiens*, discovers fossils of these plants. In 1962 the geologist Douglas Rankin and paleontologist Erling Dorf announced their discovery of plant fossils to the world. Their report in the *Journal of Paleontology* brought noted paleobotanists to Maine eager to understand the relationship of these ancient plants to those that populate our forests today.

The paleobotanists used the Doctrine of Uniformity first proposed by Charles Lyell in 1837: *Until we have evidence to the contrary, we should assume that everything in the past worked exactly the same as we see now—the present is the key to the past.* Fossil interpretation is dependent on this principle also known as *uniformitarianism.* This states that ancient geological processes occurred at the same rates in the same manner that we observe today and can account for all the Earth's geological features. Today, the Earth's history is considered to have been a slow, gradual process, punctuated by occasional natural catastrophic events.[1]

Through the work of paleontologists you can now go to the Maine State Museum in the state capital in Augusta, Maine, and see a beautiful fossilized specimen of *Pertica quadrifaria* on display. Here you would learn that, in its day, it was one of the earliest vascular plants to grow on land. Even more surprising, it started its life in a warm tropical or subtropical climate south of the equator. And you might wonder how it, as a fossil embedded in its rock strata sitting on an ancient continent, made its way from south of the equator to its present-day location in New England, i.e., Maine, thousands of miles away. Have you heard of continental drift? What changes must have occurred on Earth along its route to where it was discovered in 1968? If you could know the answers to these questions, you might come away with a different perspective on nature and the natural world. Such a perspective would convey to you some sense of the enormity of time and the events that shaped the world and life on our planet. And, if you thought about how the world is changing and how all of this relates to you, you might see your own life in a different way.

Today, scientists have shed light on these questions and thoughts. They hypothesize that the remains of *Pertica* were the result of its premature death. This, perhaps, came about because of a storm that washed tons of smothering mud and silt over it, and that over the next hundreds of thousands of years, entombed in the Earth, it was transformed into a fossil. They know that the fossil plant, now found in an identifiable rock formation, the Trout Valley Formation, is part of a massive continent of solid rock, called a tectonic plate. The North American Plate, as it is called, was one of several continental plates into which the Earth's crust had broken. The plates, however, were and are not stationary: convection currents in the Earth's mantle, like hot water rising and circling in a pot, caused them to move—some apart, some side-by-side, and some together. In those plates coming together, one plate often moved under the other in a process called "subduction" resulting in a "subduction zone"—a deep trench in the crust of the Earth. The moving plates caused volcanic activity, earthquakes, mountain building, sea-floor spreading, and other massive events.

Although *Pertica* grew, died, and was buried on the North American Plate, it was originally south of the equator in the southern hemisphere of the planet in the early part of our story. However, that was destined to change as the tectonic plate was moving northward. Today, *Pertica* is found in a special place in Maine now called Baxter State Park. There, due to erosional forces, some of *Pertica*'s remains, although fossilized, are once again bathed in sunlight. And thanks to the work of scientists in paleogeography, who reconstruct the movement of ancient tectonic plates, we can now follow *Pertica*'s 390-million-year, zigzag journey of thousands of miles across the world.

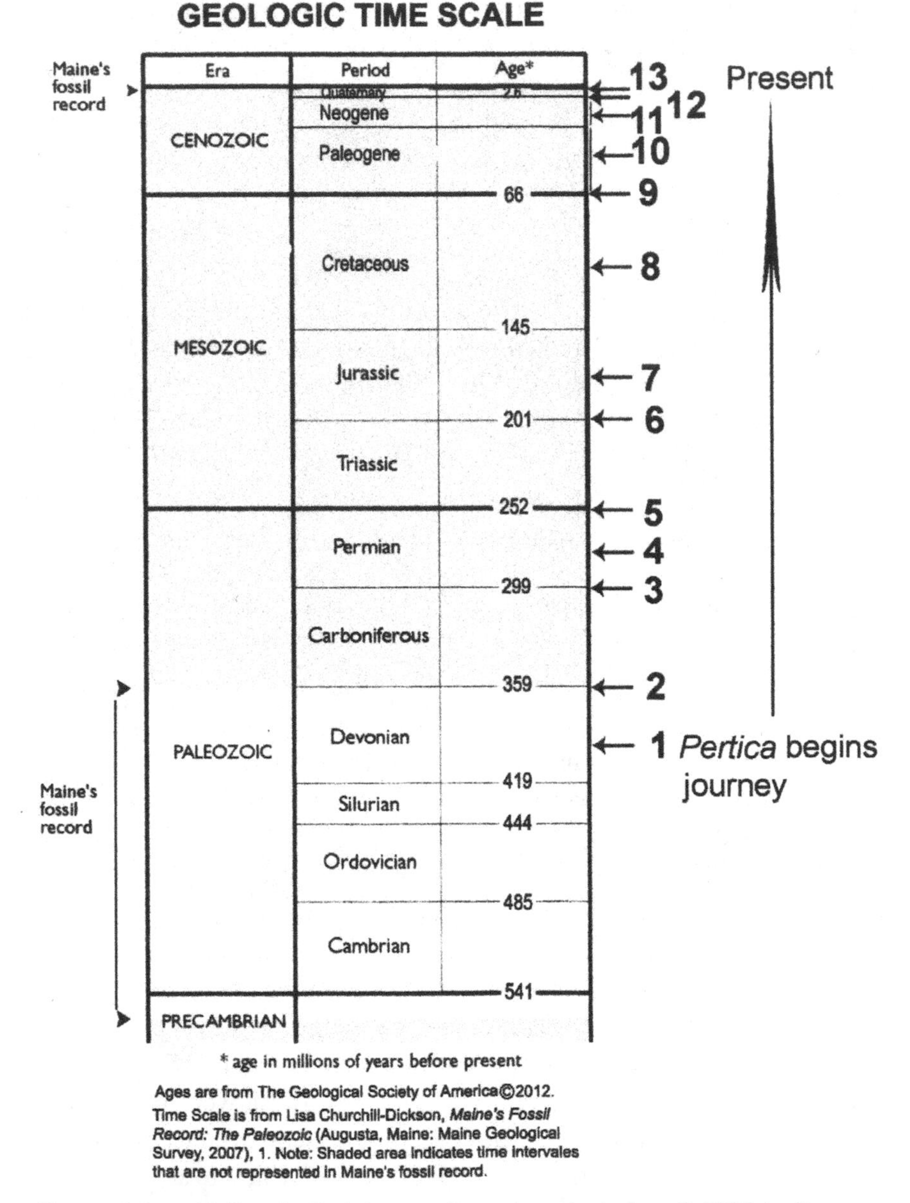

GEOLOGIC TIME SCALE
Maine's fossil record
Era
Period
Age*
CENOZOIC
Neogene
Paleogene
66
MESOZOIC
Cretaceous
145
Jurassic
201
Triassic
252
PALEOZOIC
Permian
299
Carboniferous
359
Devonian
419
Silurian
444
Ordovician
485
Cambrian
541
PRECAMBRIAN
Maine's fossil record
13
12
11
10
9
8
7
6
5
4
3
2
1 Pertica begins journey
Present
* age in millions of years before present
Ages are from The Geological Society of America©2012.
Time Scale is from Lisa Churchill-Dickson, Maine's Fossil Record: The Paleozoic (Augusta, Maine: Maine Geological Survey, 2007), 1. Note: Shaded area indicates time intervales that are not represented in Maine's fossil record.

Figures 1.1 and 1.2. *Pertica*'s journey through geologic time © 2020 by Dean and Sheila Bennett. Reconstructed by Dean B. Bennett with the help of maps produced through the work of geologist and paleogeographer Christopher R. Scotese.

PERTICA'S JOURNEY ACROSS THE GLOBE

copyright 2020 by Dean & Sheila Bennett

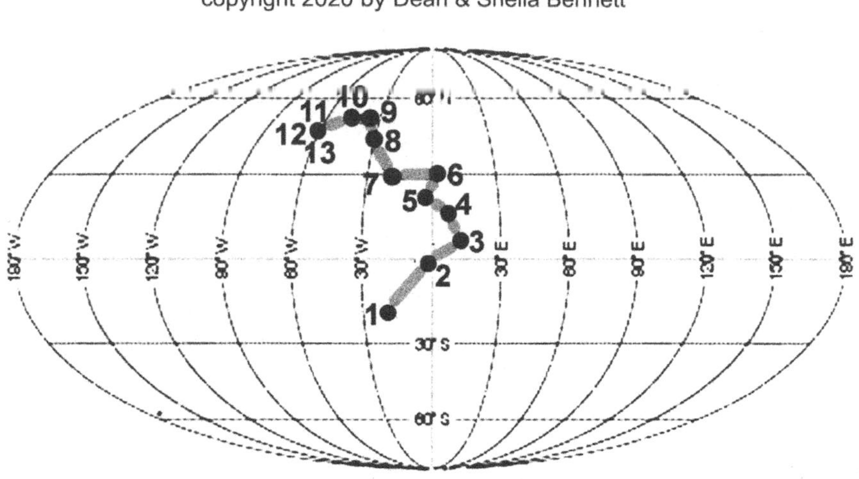

Note: Numbers refer to points of time along *Pertica*'s route.

During the fossil plant's journey, it passed through several periods of geological time, each measured in millions of years.

1. *Pertica* started its journey in the Early Devonian Period when it was buried in sediment promoting fossilization in the Trout Valley Formation. It grew on part of a moving tectonic plate some 398 million years ago (mya). Plant diversification occurred in a relative rush during the Devonian Period, a time which also saw the evolution of amphibians, animals that remained the planet's dominant vertebrate for more than 100 million years. But, it was a period of time when the world lost an estimated 75 percent of its species.
2. As *Pertica* continued to be carried along by the incredibly slow movement of the Earth's crust, it entered the Early Carboniferous Period 359 mya. Here, at about 335 mya, the tectonic plate *Pertica* was in joined other plates forming the supercontinent Pangea. Twenty million years later, 315 mya, the first reptiles arrived on the surface of its plate.

3. By the time *Pertica* had reached the end of the Late Carboniferous Period, 299 mya, conifers, to which *Pertica* is thought to be related, already had evolved.

4. The Carboniferous was followed by the Permian Period, beginning 299 mya, during which the world experienced another mass extinction, losing 96 percent of its marine species.

5. At the end of the Permian Period, 252 mya, *Pertica* entered the Triassic Period. While deeply buried in the Earth's crust of the Triassic, between 243 and 233 mya, dinosaurs evolved. They were followed by the appearance of mammals, 225 mya, also in the Triassic.

6. At the end of the Triassic Period and the beginning of the Jurassic Period, 201 mya, the world experienced yet another mass extinction with a loss of biodiversity approaching almost 80 percent.

7. On *Pertica*'s way through the Jurassic to the middle of that period, 175 mya, Pangea began breaking apart. Around 150 mya, birds appeared.

8. After *Pertica* had entered the Cretaceous Period, flowering plants came on the scene around 140 mya. Before the end of the Late Cretaceous, 94 mya, the South Atlantic Ocean opened up.

9. About 66 mya, *Pertica* came to the end of the Cretaceous and entered the beginning of the Paleogene Period, crossing the K/T boundary. It was here that a mass extinction occurred, and the Earth lost more than three-quarters of all plant and animal species. Ten million years later, 56 mya, *Pertica* reached another important time boundary between the first two epochs of the Paleogene, the Paleocene and Eocene (epochs are divisions of time within periods). This boundary, called the PETM, marks a time in Earth's history known as Paleocene-Eocene Thermal Maximum. It is a time when, for perhaps 100,000 years, the Earth experienced the highest temperatures of the last 66 million years. The temperatures were attributed to high atmospheric carbon dioxide levels.[2]

10. By 23 mya, *Pertica* was being carried through the Paleogene Period into the Neogene Period.

11. In the Neogene Period, 14 mya, *Pertica* arrived in a more modern-looking world.

12. Near the end of its journey, 2.6 mya, *Pertica* reached the Quaternary Period, and at 18,000 years ago, the world witnessed the most recent expansion of its ice sheets at the poles, which, for the previous 30 million years, had been expanding and contracting. That was when an ice sheet bulldozed across the land of Baxter State Park, exposing the fossils at the surface.

13. Finally arriving at a time humans call the present, *Pertica* was discovered and became the subject of investigation by paleobotanists.

This opportunity to follow the route of *Pertica* across the world over time is due to maps produced through the work of geologist and paleogeographer Christopher R. Scotese.[3]

Today, the fossils of *Pertica*, along with neighboring fossils of other species, exist in northern Maine's Baxter State Park, far from where they started their 8,400-mile-long journey in the Southern hemisphere, a distance from New York City to within 600 miles of the South Pole. And it is here in northern Maine that we begin our story of one of the Devonian's most distinguished fossil plants.

1

A FATEFUL MEETING

"WHAT ARE YOU DOING?"

The man in the road cut looked up. A government jeep idled at the edge of the road just above him, and he saw three men looking down at him through the vehicle's open windows.

"Looking for fossils," he replied.

"You won't find anything there," one of them said.

"Well," he said, reaching into his pocket and looking up at them with a glint of humor in his eyes, "you could go along if you want to, but I'm perfectly happy with the fossils I'm finding."[1]

Figure 1-1. William (Bill) Forbes at road cut. 1964. Photo courtesy of Warrena Forbes.

Interested, the three watched the man. His name was William Forbes, Bill to most everyone who knew him, and of those who did, none would have been surprised to see him here, for he had been doing this since he was a boy—looking for fossils. He held up a flat rock with a fossil for the men to look at. His tanned face showed his concentration as he studied the fossil, his slightly pursed lips seemingly ready to break into a smile. He was in his early thirties with curly, brown-blondish hair reaching just below his ears. He wore a light-blue, short-sleeved shirt revealing muscular tanned arms. On the wrist of one was a watch and the hand of the other held a geologist's hammer. The top of a pen poked up above his shirt pocket. Around his neck, a small magnifier dangled from a cord. He wore well-used dungarees and rugged, heavily-scuffed leather hiking boots.

Bill Forbes was born in Bingham, Maine, on March 27, 1931, the oldest of seven children. In his boyhood, the family moved to Caribou, Maine. He didn't know it at the time, but the move put him near a cluster of Maine's most significant fossil localities. It turned out to be a stroke of good fortune, for before he had become a teenager, he had found his first fossil in a Caribou gravel pit, a Devonian brachiopod, and it was the beginning of a lifelong love for science, especially for rocks and fossils. Years later he would tell a reporter: "I was just a rock hound for years as a kid. I've been interested in them since I was first aware that one rock was different from another."[2]

Figure 1-2. Henry N. Andrews, Jr., at fossil hunt in northern Maine. From the photo collection of Andrew E. Kasper, Jr.

Finding rocks with fossils in them would be Bill Forbes' avenue to becoming known and respected as a scientist. And few if any would have predicted it, for he never received a formal education beyond high school. But he had an unusual passion and talent for fossil hunting and study, a talent which, many years into the future, would be recognized in a book written by one of the three men who had stopped by the road cut to ask him what he was doing. That man, who had not met Forbes before that stop in the road cut, was Professor Henry Andrews, a nationally recognized paleontologist, who was at the time the head of the Botany Department at the University of Connecticut. In the coming years, Andrews and Forbes would work closely in the discovery of Maine's hitherto unknown fossils.

After Andrews retired, he published his benchmark book *The Fossil Hunters*. In it he would say: "The ability to locate fossils in the field is often undervalued. I have known at least three people in my career who possessed this ability to an extraordinary degree." Bill Forbes was one of the three as was Andrew Kasper, who, as we shall see, will also become a prominent figure in this story. Andrews went on to write that he worked with Forbes and Kasper "for nearly ten years in the Devonian horizons of Maine and southeastern Canada; their knack for locating new lenses of plant fossils as we walked along the outcrops often embarrassed but definitely pleased me."[3] Walter Anderson, former Maine State Geologist, told the authors many years later that Bill Forbes, indeed, had a "phenomenal sense of where fossils are—he could smell them out."[4]

Figure 1-3. Olof Nylander in his work room, Caribou High School, 1936. Photograph detail from Collections of Nylander Museum, courtesy of VintageMaineImages.com, item #10037.

Forbes was aided in many ways in his journey to become a scientist. During his boyhood years, Forbes came under the spell of Olof Nylander, one of the region's best-known naturalists.[5]

Nylander was a collector of natural history specimens and had established the

Nylander Museum in 1939 in Caribou when Forbes was eight years old. Bill Forbes loved being outside, and with the help of Nylander, he started collecting fossils and other specimens on his many trips outdoors. The old naturalist, then seventy-five, continued to influence the boy's interest in nature until his death in 1943, when Forbes was twelve.

Nylander was born in Sweden on July 14, 1864. When a boy, he became the pupil of the Swedish geologist Sven Nillson, who directed his young protégé's enthusiasm for natural history. At sixteen, Nylander came to America, eventually settling in Caribou, Maine. From there he began exploring his northern and eastern Maine surroundings, collecting specimens of fossils, rocks and minerals, plants, and animals. Eventually, he became a field collector for the United States Geological Survey (USGS) and many colleges and universities, as well as becoming the Nylander Museum's first director and curator.[6]

Bill Forbes was to follow in Nylander's footsteps—studying natural history, collecting specimens, and teaching, and like Nylander, he would become a recognized scientist. Faced with the lack of financial resources to continue his education in college after he graduated from high school in 1949, a situation common with many Maine youth at the time, Forbes had to look for work. He found a job with a well-drilling company, which gave him time to follow his major interest in life, collecting and studying fossils.[7]

Two years later in 1951, he married Warrena Bugbee, whom he had met at a high school dance. After their marriage, the couple moved to nearby Washburn where they would raise three children and live the remainder of Forbes' life. Warrena Forbes describes her husband in those early years as around five-foot ten inches tall, slightly built, physically fit, and very strong and agile. He had blond curly hair and was considered, especially by her, as handsome. He was an extrovert, she said, and would talk to everyone. She recalls that his interest in rocks and fossils extended to anything to do with nature—plants, trees, stars, animals, and on and on. He was an avid reader but usually read scientific and historical books.[8]

After his marriage to Warrena, Forbes found employment as a salesman with the W. S. Emerson Co., a wholesale dry goods business

Figure 1-4. William (Bill) and Warrena Forbes in their Washburn, Maine, home where they met with many visiting paleontologists. Photo courtesy of Warrena Forbes.

in Bangor. Later, he worked for the National Cash Register Company. Sometime in the first half of the 1960s, he took over the ownership and running of a retail dry goods store in Washburn, The Village Shoppe. With no lessening of his passion for fossil hunting and study through these years, he continued to work on specimens between customers and searched for and studied them in evenings and on weekends.[9] Warrena recalled to a reporter that "he had the house full of rocks as well as the basement and garage. We couldn't even get the car in. He actually had tons of material."[10] In an interview after his death, Warrena told the story of celebrating an anniversary with her husband at a restaurant across the river from Caribou in New Brunswick. Having to wait for a table to be readied, Forbes noticed a nearby outcrop and expressed an interest in examining it. Warrena, wearing high heels, accompanied him to the outcrop, and while they were there, one of her high heels broke, a minor distraction from a wedding anniversary with a husband who enlivened her life with a passion for science and the outdoors.[11]

Later in his life, the scientific community recognized Forbes for beginning a detailed analysis of Aroostook County's geologic history.[12] But in those early years, he was often frustrated because of difficulties in finding information and people to whom he could turn for help in pursuit of his passion. That changed for him in the early 1960s when a number of prominent geologists and paleontologists, interested in

Figure 1-5. Ely Mencher on a geological survey in northern Maine. From the photo collection of Andrew E. Kasper, Jr.

investigating and mapping the geology of northern Maine, gradually became aware of his fossil collections and the scientific knowledge he had of the area's geology.

Forbes' association with those professional geologists and paleontologists began when Ely Mencher, from the Massachusetts Institute of Technology (MIT), hired him in the summer of 1962 to assist him and his students in investigating and mapping the Ordovician, Silurian, and Devonian rocks of northern Maine.[13] Years later, Warrrena recalled that, as a result of this job, she and her husband "became friends with Mencher and his wife and son."[14] In 1964, because of his discovery of fossil locations, collections, and knowledge, Forbes was identified as one of the authors of the paper *Reconnaissance Bedrock Geology of the Presque Isle Quadrangle Maine*, published by the Maine Geological Survey.[15]

That same year, 1964, Forbes was acknowledged by James Schopf, a paleontologist with the U. S. Geological Survey, in another paper, *Middle Devonian Plant Fossils from Northern Maine*. In his report, Schopf recognized Forbes by naming a new fossil plant after him and had written: "It is a pleasure," he wrote, "to name this species [*Calamophyton forbesii*] for Mr. W. H. Forbes of Washburn, Maine, whose enthusiasm, knowledge, and interest in the geology of northern Maine is appreciated, and whose assistance in the field is gratefully acknowledged."[16]

In addition to Forbes' knowledge and expertise, the professional geologists and paleontologists with whom he worked discovered something else about him—his generosity. He would readily share his knowledge, give them specimens he had collected, and show them the fossil sites that he knew. In the words of one scientist: With Bill Forbes there was "no turf peeing."[17]

It was also in the summer of 1964, that Forbes had that chance meeting at the road cut with three leading scientists studying the region. In the jeep were Ely Mencher, James M. Schopf, and Henry Andrews. It seemed that Mencher had invited Schopf to join his team that summer to help determine the age of rocks he was finding through fossil evidence, and Schopf, in turn, had invited his good friend Andrews to join the group. The road-cut meeting led to an invitation from Forbes to Schopf and Andrews to visit his home. After they had arrived, he took

Figure 1-6. James M. Schopf hunting for fossils in northern Maine. From the photo collection of Andrew E. Kasper, Jr.

them to the basement where he kept his fossil collections and gave them the opportunity to examine some large specimens of fossils. The two scientists were stunned: "Bill Forbes was finding some of the finest fossils available in North America and they were dealing with no amateur." Forbes said that one of them told me "that I had enough stuff in my basement to keep a crew busy for years."[18]

Andrews, in particular, was struck by what he saw. Years later, he and others recalled in a paper that "the specimens contained some spectacular fossil plants, described later as *Psilophyton forbesii* [after Bill Forbes who had collected them]. . . . It was immediately evident that here was the kind of lead paleontologists long for. . . . It was no secret that the forested areas of north-central Maine were not generally regarded as likely places to look for fossil plants."[19] Yet, here were fossils from one of those places, and it would be Bill Forbes who would lead Henry Andrews to it and the beginning of "several summers of intensive field work." [20]

That place was in northern Baxter State Park where, ten years earlier in 1955, fossils had been discovered in an unusual, enigmatic, geological formation containing some of the finest, early, land-plant fossils in the world. The discoverer was Douglas Whiting Rankin.

2

WADING INTO THE DEVONIAN

WE GO BACK TO A DAY, NEARLY TEN YEARS BEFORE FORBES' FATEFUL meeting in the road cut, a day when the first fossils in our story were discovered by Douglas Rankin in northern Maine's Trout Brook valley, the place of our story. Rankin was twenty-three years old in 1955 when he was beginning a Harvard graduate study of the geology of this area in Baxter State Park. But the rocks, minerals, and geologic formations that he needed to see, understand, and map—the foundation of the land here—were covered by a dark cloak of boreal forest, nearly impenetrable in

Figure 2-1. Douglas Whiting Rankin at age 21 when he graduated from Colgate University. That fall he entered Harvard University for graduate work in geology. Photo of Douglas Rankin from *Salmagundi,* Colgate University 1953 student yearbook, Special Collections and University Archives, Colgate University Libraries. Published by permission.

places. Brooks and streams, however, had made avenues of discovery by removing the top soil and its plants, and by eroding into the underlying rocky surface of the land, exposing its secrets to those with trained eyes, like Rankin.

From the South Branch Pond's outlet, he would have seen the brook flowing through a sandy channel into a small, alder-crowded marsh below, and he would have found the water chillingly cold, even in late summer, not unusual in northern Maine. If he had hoped for a dry spell of low water for easy wading, he would have been disappointed. That had occurred four months earlier in April, and now at the end of summer, the water level was up: the month of August had been the rainiest month of the year, about four-and-one-half inches—an inch more than usual. Still, there was no doubt that he was going to walk the three miles down the brook anyway; this was his task, what he liked to do. It was his future, and before the day ended, he would make two discoveries—important discoveries.

What Rankin found would influence the lives of others as well as his own. It would bring him back to this place many times through the years to lead others to it, and they, too, would be as surprised and entranced as he had been. The discovery would catch the interest and curiosity of other scientists who would explore the area. One, in particular, would make his career based on what he found—a finding that would capture the attention of the public and lead to legislative halls and a law from which would come a new symbol for the State of Maine. But it wouldn't stop there: scientists would be drawn back here time and time again right up to the present where today a new generation continues to explore, study, and contemplate the history of our planet, its future and ours as they make new discoveries here. But back in 1955, as Doug Rankin prepared to wade down South Branch Ponds Brook, this would all come later.

It is quite possible that before Rankin stepped into the cold waters of that mountain stream he had hiked up onto North Traveler Ridge overlooking this country of the South Branch Ponds. Here, only two years before, a new trail had been completed, which afforded a spectacular view. From the ridge he could have seen the north end of Lower South Branch

Figure 2-2a. Lower South Branch Pond from North Traveler Ridge. Rankin started down the stream at the pond's outlet at the right end of the pond. Photo by Dean B. Bennett.

Pond and in the distance the pond's outlet. It would have been easy for him to pick out the break in the trees where the brook flowed from the pond, passing Little Peaked Mountain to the east and by the mouth of Gifford Brook on its way to Trout Brook in the distance. Beyond, he could have made out the low outline of Wadleigh Mountain on the north side of Trout Brook valley. This valley, lying between Wadleigh Mountain and North Traveler Ridge, would be written about, read about, and talked about in certain circles for years to come.

The valley of Trout Brook is surrounded by mountains. Two mountains bracket its east and west ends. On the western end near the confluence of Trout Brook's North and South Branches, Burnt Mountain rises 1,793 feet. At the valley's eastern end lies the 1,769-foot-high Trout Brook Mountain near where Trout Brook flows into Grand Lake Matagamon. South of Trout Brook Mountain just east of the South Branch Ponds, the Traveler stands at 3,541 feet, twice the elevation of either Burnt or Trout Brook Mountain. Farther south, about fifteen miles

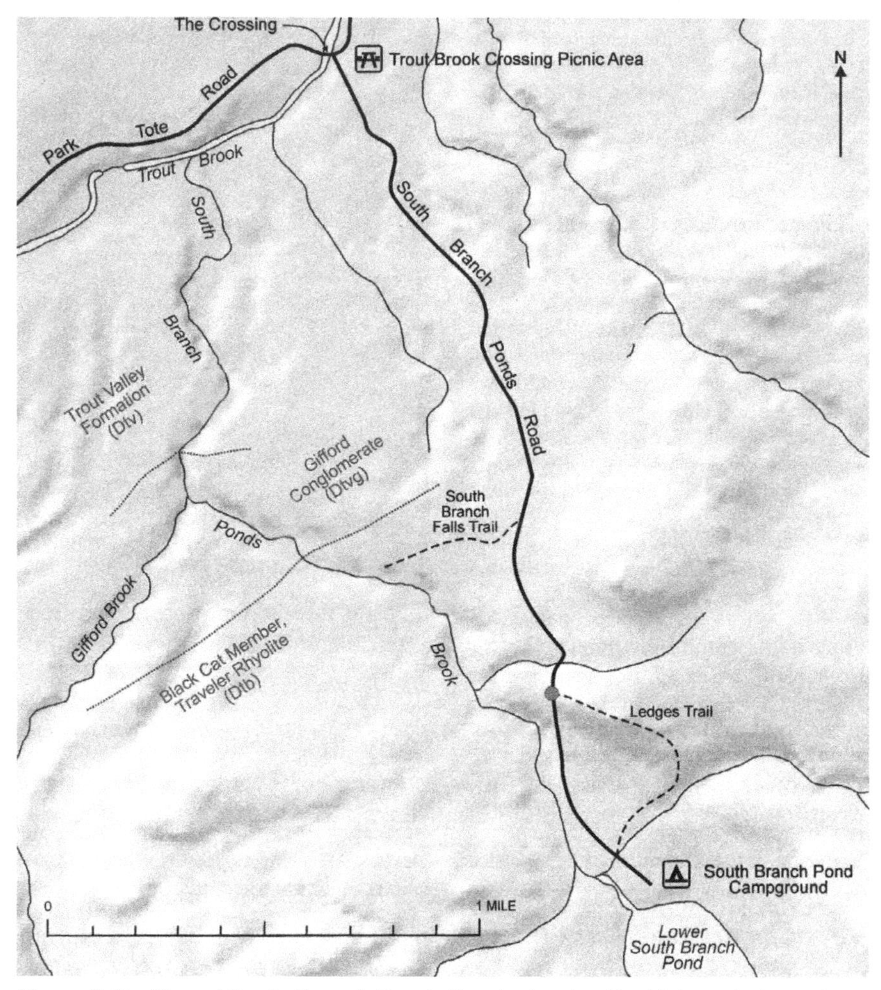

Figure 2-2b. Map of South Branch Ponds Brook showing Rankin's route from the mouth of South Branch Pond to Trout Brook. Adapted and published by permission of the Maine Geological Survey, Department of Conservation in *A Guide to the Geology of Baxter State Park*, 2010, p. 67, by Douglas W. Rankin and Dabney W. Caldwell.

from North Traveler Ridge, Baxter Peak on Mount Katahdin looks down on it all.

Doug Rankin would later say that his three-mile hike down South Branch Ponds Brook to Trout Brook and to where it is crossed by the

Figure 2-3. The authors wading down South Branch Ponds Brook, August 20, 2016, following in the footsteps of Douglas W. Rankin when he first studied the mountain stream in 1955. It was then that he discovered the Trout Valley Formation and its early Devonian plant fossils. Photo courtesy of Cheryl B. and Charles O. Martin.

Park Tote Road (called The Crossing) was "one of the most geologically rewarding in the Park."[1] When he began his trek at the outlet of South-Branch Pond, Rankin slogged through a shallow, marshy area thick with alders. Soon he detected a current in the swale, followed by the sound of ripples ahead as the boggy area narrowed to a brook. He moved out of the stream up onto its bank and better footing. About a mile from the pond he came to a low flat ledge just at stream level. Here he saw a mysterious pattern of intersecting concentric joints appearing to curve about a center. Nearby, above the stream on its east bank, he also saw columns of gray rock of varied diameters, but strangely each had multiple sides.[2]

Later, after further examination, Rankin would identify these outcrops as being of the "Black Cat Member" of the Traveler Rhyolite, named after a nearby mountain. Rhyolite is an extrusive igneous rock,

Figure 2-4. Columns of Traveler Rhyolite. South Branch Ponds Brook. Photo by Dean B. Bennett on 1994 field trip led by Douglas W. Rankin.

meaning that it flowed as magma (molten rock) on the surface of the Earth or was ejected as ash from a volcanic vent in the earth's crust before it cooled. It is the equivalent of granite, which formed from magma underground and of which Katahdin is formed.

For now, however, a closer look with a hand lens enabled him to identify the rock as "tuff." Tuff is formed of volcanic ash from an exploding volcano or from being emitted from a volcanic vent. In this case, the tuff was composed of glass-rich materials from a highly heated mixture of volcanic gases and ash flowing down the flanks of a volcano or along the surface of the ground. The materials had been laid down in layers which, over time, were compacted, welded together, and hardened by heat and pressure from overlying materials. Rankin identified this as "welded ash-flow tuff."

Many of the rocks that make up the geological character of an area have layers (foliations) that form flat surfaces. While some rocks are igneous, that is, formed from magma, which is molten or partly molten material, such as rhyolite and granite, others are sedimentary in origin, that is, composed of sedimentary beds of silts, clays, sands, or other particle material originally deposited in horizontal layers. Because the tuff Rankin saw is deposited from volcanic ash and laid down in layers, it has the layering characteristic of sedimentary rock. However, over time most layers or beds of sediments are no longer horizontal due to tectonic (earthmoving) forces which tilt, fold, or otherwise change their orientation. Rankin's task was to measure the orientation of the layered rock structures in the outcrops.

Two measurements were crucial to this task—the strike and the dip. The *strike* is the compass direction of the surface of a bedding plane of sedimentary or stratified rock (rock composed of layers) as it intersects the surface of the Earth. The compass is also used to measure the *dip*, which is the angle that the surface of such bedding plane makes from level or horizontal. Before he left the site, Rankin took out his compass and measured the strike and dip of the outcrop with its layers of ash flows. In his notebook, he recorded the strike as N80°E and the dip as 30°N., figures he would use in descriptions of field trips he would lead here into the 1990s. He also recorded them at this location on a map.

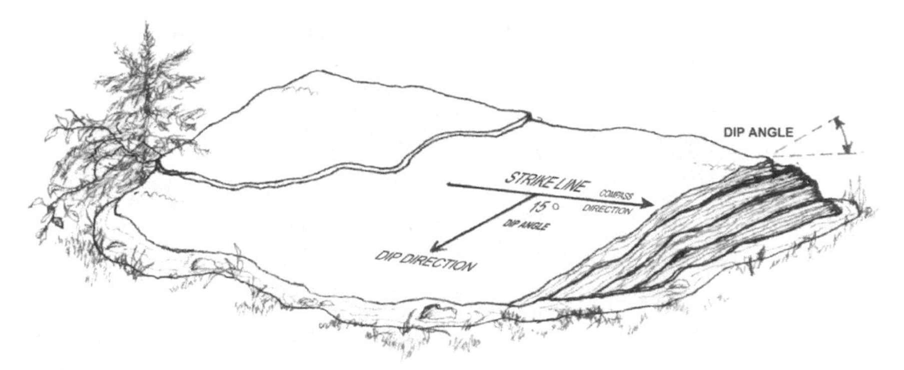

Figure 2-5. Strike and dip diagram by Dean B. Bennett.

A few hundred feet downstream, Rankin stopped to examine a series of low ledges lying crosswise of the stream. Once again, these ledges were exposures of the Black Cat Member of the Traveler Rhyolite—jointed (fractured) welded ash-flow tuffs that heat and pressure had hardened and consolidated into a glassy texture. The ledges here also dipped, or slanted, about 30° N., with the layers of solidified ash flows striking in a N70°E. direction. He also saw crude columns standing upright, nearly perpendicular to the ledge surfaces.

A relatively short distance from the low ledge outcrops, Rankin came to the falls of the brook. Here, the quiet rippling stream he had been following turned wild as it tumbled and roared over and around ledges of the Traveler Rhyolite, where hundreds of millions of years ago unimaginably violent forces had formed them of volcanic ash and left them upturned. Those who would follow in Rankin's footsteps in later years would find his measurements for this location published on a 2010 map with the geological symbol for strike and dip. And for those who would have the good fortune to accompany him here, he would always point out that if you "are easily distracted from geological wonders," the pools at the bottom of the falls "offer excellent swimming."

Below the falls and pools, he found easy walking in the woods for about a quarter of a mile on the east side of the brook before crossing it. He didn't know what geological surprises lay ahead of him. Entering the brook again, he saw ledges of rounded, cemented pebbles and cobble-sized rocks of the Traveler Rhyolite making up a kind of rock he knew as conglomerate. Conglomerate is a sedimentary rock composed of rounded and variously shaped rock fragments called clasts. This particular conglomerate eventually would become known as Gifford Conglomerate after the nearby Gifford Brook. One ledge of the conglomerate that caught Rankin's eye was a gray, flattened ledge in the stream's bed, a feature geologists call "pavement" because of its similarity to the pavement of a road. He also observed that on the steep bank above the low ledge the conglomerate was covered by glacial till, the name given to a mixture of sand, gravel, pebbles, and larger rocks deposited by the last glacier thousands of years ago during a time called the Pleistocene. Above the till, a ragged strip of topsoil left by the ravages of the spring's high water supported the edge of the forest encroaching on the stream.

Figure 2-6. Gifford Conglomerate caves formed by erosion from South Branch Ponds Brook. Photo by Dean B. Bennett.

About 200 paces downstream, Rankin encountered another outcrop of Gifford Conglomerate forming a bank on the east side of the brook. Its whole surface had a coarse, knobby or bumpy texture formed by rounded rocks (clasts) of pebble and cobble size, brown-to-orange-hued in color. Planar factures in the conglomerate which cut across the cobbles gave him clear evidence that the rounded rocks of rhyolite in the conglomerate were cemented as solid rock before the breaks occurred. He could tell by their rounded and smoothed corners and polished surfaces that the conglomerate's pebbles and other rocks had once been transported, bumped, shaped along a streambed by running water. From the angle of the conglomerate's beds, it was also evident to him that they had been tilted slightly toward the north.

Continuing down the brook another 200 yards, he came to a canyon wall of conglomerate that he estimated to be more than thirty feet high. In the wall near stream level, his attention was caught by a fascinating set of caves eroded by the brook leaving an amazing scene of sculptured outcrops of conglomerate. Near the top of the wall he saw sandstone beds interlayered with the conglomerate. The conglomerate was cut by vertically oriented joints. All in all, the features guaranteed the presentation of a strange and interesting scene to visitors who stumbled upon them.

A short distance beyond the canyon wall of conglomerate, Rankin came to Gifford Brook where it entered South Branch Ponds Brook from the southwest. The brook flowed swiftly from a narrow opening in the thick woods, tumbling over a rocky bed of boulders alongside a shadowed rock wall eight or more feet high studded with rocks cemented in conglomerate. The wall continued down the west side of South Branch Ponds Brook for a short ways.

In another few hundred feet, Rankin came to a high ledge on the left, or west side, of the stream, no longer a body of coarse conglomerate. He determined that by the upward sequence of bedding layers he was above the conglomerate and that he was now into what he believed to be the beginning of the main body of a large unnamed rock formation in the valley of Trout Brook. The ledge contained numerous lens-shaped rock bodies of hard, dense, black quartz that he identified as chert. One of these was later believed to be part of a fossil called *Protaxites*. Although

Figure 2-7. Mouth of Gifford Brook at junction with South Branch Ponds Brook. Photo by Dean B. Bennett.

its identification would be long debated, he concluded that it possessed characteristics of algae, although, today, others believe that it is more like a fungus.

Farther downstream in the unnamed formation in a layer of shale rich in carbon, giving it a black, coal-like color he made what was perhaps his most significant finding. It was his first discovery of plant fossils, a discovery that had never been reported. This was an unusual find here in this area of northern Baxter State Park, an area heavily dominated by granite, rhyolite, and other igneous rocks that had undergone extensive geological change by earth movement, heat, and pressure which could destroy evidence of early life. Yet, here was a place where Rankin would make his first collection of plant fossils.

Although the layer of shale was of unknown extent, the fossils indicated that somewhere, if not here at this spot, ancient plants had once grown and had been buried in sediments. The plant fossils were fragmented, broken up, probably before or during transportation by flowing water. They were impressions of plant stems, poorly preserved in black

Figure 2-8. Sheila K. Bennett examines a lens of shale containing Devonian plant fossils on South Branch Ponds Brook near Rankin's discovery of them. Photo by Dean B. Bennett.

sandy shale. Impression fossils are similar to compression fossils in that both are two-dimensional. Impression fossils, however, do not contain organic material, being essentially an imprint in fine-grained sediment made, in this case, by a plant which eventually decayed leaving an impression to be fossilized.

Undoubtedly, Rankin's discoveries created a sense of anticipation as he approached every turn of the brook. At one point, he came across a fine-grained rock called diorite that had intruded into a break or fracture forming what is known as a dike. Observing that it cut across the bedrock, Rankin knew that the dike was younger. In fact, it is now thought to possibly be the youngest rock in Baxter State Park.

As Rankin continued down the brook, he discovered "scraps of plant fossils in nearly every outcrop of sedimentary rock." It was during the summers of 1955 to 1957 in the vicinity of this area that he first discovered an exposure of black shale containing fossil specimens of the plant *Psilophyton* with spiny stems. *Psilophyton* is a genus of extinct vascular plants of Devonian age (about 420 to 360 million years ago) first described in 1859. It was here in August of 1959 that Professor Erling Dorf of Princeton and William Forbes visited the site with Rankin, and it was also here that Forbes collected smooth-stemmed *Psilophyton* specimens, which were eventually named *Psilophyton forbesii* after him.

When Doug Rankin finally waded out of South Branch Ponds Brook that day in 1955, he knew that he had discovered something different, and in the years to come, he would penetrate the area's remoteness, crawl beneath its thick cover of forest, study the clues in its rocks, and come to understand more clearly how unique it is in the world. And so would many others, for he would map it and lead trips here throughout his life, helping them to discover a passion to explore the wonders of the earth—a passion that had begun in his boyhood.

3

PREPARING TO PURSUE A PASSION

DOUGLAS WHITING RANKIN WAS BORN ON SEPTEMBER 9, 1931, IN Wilmington, Delaware, into a professional family. His mother, Helen Barnes Whiting, graduated from Mount Holyoke College in 1915 and received an MA from Brown University in 1918. She was an instructor at Brown University until 1920.[1] Rankin's father, Carl Seib Rankin, received a civil engineering degree from Lafayette College in 1911 and a degree from Columbia University in 1912. Later, he joined the faculty of the Department of Engineering at the University of Delaware. By 1925, he was an assistant professor of engineering and mathematics.[2]

Rankin's parents were outdoor oriented, and from their vacation-home-base at Lake Fairlee in Vermont near the New Hampshire border and only fifteen miles from the White Mountain National Forest, they introduced Rankin at age five to the White Mountains.[3] From that age through the early years of his youth, his parents took him and his older brother, Bruce, on many family hiking and camping trips into the rugged terrain of the Granite State. It was there that his love for nature and the outdoors was developed and nurtured—a place where thousands of years of glaciation, weather, and water had exposed spectacular displays of rocks and minerals and sculptured, smoothed, and polished them into infinite forms and shapes. Not surprisingly, it was a place where Rankin developed an early interest in geology.[4] But it was more than an interest for him: it developed into a curiosity about the natural world that led him, ultimately, to taking his first steps into a life of science.

Figure 3-1. Douglas Rankin (center) in 1953 as an Appalachian Mountain Club (AMC) trail maintenance crew leader. From Milne Special Collections and Archives Department, University of New Hampshire Library, Durham, NH. Published by permission.

In the summer of 1949, Rankin's love for the outdoor life and adventure took another turn when he began working on a trail crew for the Appalachian Mountain Club (AMC). More than sixty years later, he reflected on this experience: "In those unreconstructed days, the crew consisted of about ten high school or college-age males . . . we worked ten six-day weeks. . . . The Appalachian Mountain Club had made patrolling and 'standardizing' trails a priority since the early 1920s. We were also responsible for the upkeep of shelters, toilets, can pits (this was pre-'carry in, carry out' policy), ladders, and bridges."[5]

By chance, Rankin's crew was assigned to Baxter State Park's Russell Pond Campground in northern Maine.[6] This was a time when the AMC, earlier in that decade, had worked out an agreement with park officials to maintain many trails on Katahdin.[7] The Russell Pond Campground is one of the most remote in the park. It required more than a seven-mile hike to reach it from any direction. From the campground, trails provide access to several other backcountry campsites, including Wassataquoik Lake and Wassataquoik Stream. The campground is located on a rocky, shallow, twenty-acre pond with good views of the surrounding mountains. It is not uncommon to have a moose walk by one's lean-to at night or to meet up with one on a trail. For Rankin, the experience at Russell Pond Campground developed into a life-long interest in the park and the wild land surrounding it.[8]

At the end of his summer's work on the AMC trail crew, Rankin entered Colgate University in Hamilton, New York. There he majored in geology. He was fortunate to have two professors who not only instilled in him a passion for science and geology but prepared him for a lifelong future in the field. Professor John Grant Woodruff had a reputation as a skilled geologist and a gifted instructor and was remembered as being "highly effective as a teacher of field geology" (an area in which Rankin would excel during his career). Although Woodruff's courses were demanding and had high standards, they were especially popular among undergraduate students and instrumental in attracting hundreds of new geology majors.[9] Rankin's other professor was David Woolsey Trainer, Jr., remembered for his sense of humor, his observation skills, and his enthusiasm for geology. As a graduate student and instructor at Cornell University, Trainer was highly regarded by his associates. His friendliness and cheerful manner endeared him to both faculty and students.[10]

During his four years of undergraduate work, Rankin made the dean's list each year. He played soccer, joined the Sailing Club, and was active in the Outing Club.[11] During his final year at Colgate, he served as president of the Outing Club. The club provided opportunities for outdoor life, including hiking, camping, hunting, and skiing. That year, in order to provide a greater opportunity for underclassmen to participate, the club bought a used hearse for transportation. They made good use of it for they climbed Mt. Marcy, the highest mountain in the Adirondacks, explored caves in New York state's Howe Cavern region, and, that winter, made several skiing trips. The school's yearbook for that year reported that, under Rankin's leadership and that of a faculty member who was the club's winter sports advisor, "the Club had its most successful season in years."[12]

Meanwhile, Rankin continued working on the AMC's trail crews each summer.[13] This, by itself, spoke of a young man who loved the outdoors, who was not afraid of hard work, who set goals and persevered in their accomplishment, who enjoyed working with others, and who believed in working for something greater than himself—all characteristics that would lead him to the success he found in his future career. But, for a moment, in the spring of 1953, fresh from his graduation at Colgate with a Bachelor of Arts degree cum laude and having been accepted by Harvard University to begin graduate studies in geology in the fall, he faced a dilemma. He expressed it years later in an article: "I had just completed four wonderful years on the trail crew but knew that, for my career, I should seek geological work the next summer." But he really wanted to work for AMC that final summer as a trail master of a crew with the major responsibility to build a suspension bridge. It was needed to replace one that had washed out over the West Branch of the Peabody River in New Hampshire. The Peabody drains the Great Gulf, a huge glacial cirque (a steep mountain basin) on the north side of Mount Washington. As far as he knew, no trail crew had constructed a bridge since the Great Hurricane of 1938. "The challenge of the Great Gulf Bridge won," he wrote. At his request, his engineer father designed the bridge and Rankin and the crew successfully completed the complicated construction by the end of the work season on September first. It was a major accomplishment for a trail crew, and one, Rankin learned years later, which some in the AMC had not thought possible for unskilled college students.[14]

There was, however, another event that had a much greater significance in his life during that summer. It occurred while he was having dinner with his father and brother in Pinkham Notch during the bridge planning process. He saw Mary Backus for the first time that season.

Figure 3-2. Douglas W. Rankin in 1953 sitting on the Appalachian Mountain Club's (AMC) West Branch Peabody River bridge that he and his crew built. Published by permission from Douglas W. Rankin Family. Photo by Scott Southworth.

She was working her second summer on the Pinkham crew. "Because of the proximity of the bridge site to Pinkham," he later wrote, "I saw Mary numerous times during the summer."[15] Three years later they married, "and together they developed a shared love of New England geology that lasted a lifetime."[16]

In the fall of 1953, Rankin graduated from Colgate, receiving a Bachelor of Arts degree cum laude in geology. He immediately began his graduate studies in geology at Harvard, where he would excel in mineralogy and petrology (the study of rocks). After receiving a master's degree, he began work on his PhD thesis under the supervision of professors Marland P. Billings and James B. Thompson, Jr. Aware of Rankin's interest in studying igneous rocks and working in the field where he loved to be close to the natural world, Billings and Thompson directed him to study and map the rhyolites and granites of the Katahdin area in northern Baxter State Park. It was an opportunity to unravel the igneous geology of an area not yet fully understood. The challenge excited him, for it would take him to the remote mountains, woods, and waters of a place he had come to love in his earlier trail work with the AMC. It was a place where he could truly follow his passion, and it would take him ever more deeply into the Maine woods in the footsteps of others, some of whom had been equally as curious about what shaped this landscape and the foundation upon which it was built.

4

FACING A PEOPLED WILDERNESS

When Doug Rankin arrived in northern Maine to begin his PhD work in 1955, he faced a daunting task by any measure. His study area covered 300 square miles, extending from Katahdin's mile-high Baxter Peak on the southern boundary of the area to about three miles beyond Grand Lake Matagamon on the northern boundary and from Nesowadnehunk Lake on the west side to Bowlin Pond on the east side. Rugged mountains cover a large part of the area, which contains three of the six highest mountains in Maine. The northern part of the study area is drained by the East Branch of the Penobscot River and its tributaries, including Trout Brook, which figures prominently in this story. Trout Brook flows along the northern edge of the area's mountains into Grand Lake Matagamon, the largest lake in his study area. The western and southern slopes of the mountains in the area lie in the watershed of the West Branch of the Penobscot River. The higher mountains in the southern half of the study area are of granite, and the lower mountains in the northern part of the area are composed of rhyolite. Although the area had undergone heavy logging and devastation by three major forest fires by the time Rankin arrived, much of it had once again become densely wooded with coniferous and deciduous trees.

"There are no towns, settlements, active farms or paved roads within the area." Rankin wrote. "The only through road is the dirt road from Patten to Greenville." Drivable roads in the area totaled only sixty-five miles and made general access difficult, but, he pointed out, the wilderness is broken by "many old tote roads, decaying dams, and stumps

Figure 4-1. Douglas Rankin's 300 sq. mi. study area from Katahdin to Grand Matagamon Lake. Seen from a scenic overlook in Patten, Maine. Photo by Cheryl B. and Charles O. Martin.

throughout the area" as well as the remains of old hay farms. The two nearest towns are twenty to thirty miles away.[1] There were, however, some recent changes that worked to Rankin's advantage when he arrived on the scene. A major road along Trout Brook, called the Tote Road, had been "newly constructed" ten years earlier. It ran from an old hay farm at Grand Lake Matagamon to a place called The Crossing where a side road crossed Trout Brook and provided access to the South Branch Ponds. This road had been rough and usable only by horse-drawn wagons until 1951, four years before Rankin's arrival. And only three years before, the campground at South Branch Pond with lean-tos and a ranger's cabin had been established. Roads from this area to the southern part of the park at Sourdnahunk Field were not passable until just before his arrival, sometime between 1950 and 1955.[2] Some of the park's trails in Rankin's study area were also not in existence in 1955. In fact, during his 1955 field

Figure 4-2. Interior Baxter State Park. Photo by Dean B. Bennett.

season, he assisted an AMC trail crew by scouting out and arranging for the crew to cut out a trail on the upper part of South Turner Mountain.[3]

The heavily forested land Rankin saw when he arrived belied the enormous natural forces and severe human-made changes it had experienced. After the Laurentide Ice Sheet had covered the land for millennia beneath a blanket of ice a mile or more thick, it began to retreat about 22,000 years ago. By about 14,000 years ago, the ice had retreated onto land in the upper reaches of the East Branch of the Penobscot River on the eastern boundary of Rankin's research area. By about 11,000 years ago, most of it was gone.[4]

Through the thousands of years that passed following deglaciation, climate changes and plant succession brought about a mixture of vegetational patterns ranging from tundra (which still exists at high elevations) to woodland forests that are predominant today. Animal life followed and varied with changing habitats. Meanwhile, humans left a light footprint on Rankin's study area, until Europeans arrived in the 1600s. Still, it wasn't until 1820, the year Maine became a state, that events were underway that would radically change the face of the land. In America the first efficient mechanized papermaking plant had been in production for a few years,[5] and by this time, too, the practice of log driving had already been instituted.[6] In Maine another means of transporting the state's wood resource was beginning to change the landscape: roads, especially what is known as the tote road. One that would have a critical effect on Rankin's area was a tote road from Patten to the East Branch of the Penobscot River.[7] This came at a time when Massachusetts and Maine were poised to begin granting and selling millions of acres of public lands. But before that could be done, surveying became a necessary step, and out of it came the boundaries of Rankin's study area.

Between 1825 and 1933, surveyors established a baseline, called the Monument Line, for laying out townships in northern Maine's public lands. The line began in the remote headwaters of the St. Croix River in eastern Maine and ran westward ninety miles, crossing the East Branch three-quarters of a mile above its confluence with the Sebois River and continuing through Katahdin's Northwest Basin.[8] North and south from this line, mapmakers and surveyors began laying out townships in squares, six miles on a side or thirty-six square miles per town. Each town received two numbers: a "row number" for the row of towns in which it was located, all towns in the same row receiving the same number, and a "range number" for the column of towns in which it was in. For example, one of the key townships in Rankin's study area, in which the Traveler Mountains, South Branch Ponds Brook, and parts of Trout Brook are located, is T5 R9 W.E.L.S., meaning a town in the fifth row and the ninth range column. W.E.L.S. means west of the eastern line of the state where the Monument Line began in the St. Croix River area.

With the land in Rankin's area surveyed and land sales underway, logging began in the 1830s along both sides of the East Branch of the Penobscot, Wassataquoik Stream to the west and the Sebois River to the east, with the logs floated down the river to the mills above Bangor.[9] Log driving would continue on the East Branch until 1946, nine years before Rankin arrived.[10] By 1840, one-third of the original pine forest in Maine had been cut.[11] In 1857, white pine was being cut on Trout Brook land.[12]

Meanwhile, private interests had gained legal control over publicly navigable rivers and streams to collectively improve them by damming and clearing them for river driving.[13] The lumbermen penetrated the headwaters of the East Branch of the Penobscot into Allagash country, and in 1841 work commenced on the building of dams and the construction of the Telos Cut or Canal. This allowed Allagash logs to be driven down the East Branch.[14] By the 1870s, the white pine phase of lumbering was over, and lumbermen and mill owners were cutting the more plentiful spruce for sawlogs.[15] Logging roads and tote roads crisscrossed the whole area, and some, such as the one along Trout Brook, became permanent.

One of the old hay farms Rankin mentioned in describing the country when he arrived was Trout Brook Farm. He found that, at least, one building, its old farmhouse, still survived, and it would become important to him when he conducted his field studies. The land was first cleared and buildings erected around 1837, and for many years it was run by David Pingree and E. S. Coe. The farm supported their lumbering efforts by providing hay, oats, and other produce; summering horses and raising smaller farm animals; and supplying tools, equipment, and necessary goods.[16] In the 1880s, it was reported that the farm consisted of "four hundred acres of cleared land, four houses, and eight or ten barns."[17]

During the nineteenth century and into the twentieth century, from the time it was built, the farm was well situated to meet the needs of the many sportsmen, adventurers, artists, photographers, writers, and others who came into the East Branch region looking for the wild character, natural beauty, remoteness, and outdoor adventure to be found here. In their wake, they left books, articles, photographs, paintings, sketches, collected specimens, and reports, many of which included references to the farm. By the turn of the century, the 1899–1900 edition of *Carleton's Pathfinder and Gazetteer* advertised the farm as a place to spend a vacation.

Among those canoeing down the East Branch in those early years were the famous ornithologist John James Audubon in 1832, and Hudson River School artist Frederick Church in 1852. Henry David Thoreau paddled the East Branch in 1857, and recorded his experience in his famous book, *The Maine Woods*. According to Rankin, Thoreau made the

Figure 4-3. Trout Brook Farm. From Myron H. Avery Collection, Maine State Library.

first recorded geologic correlation in noting the similarity in appearance between Horse Mountain and Mt. Kineo.[18] Thoreau has been called one of the first conservationists, and his writings on his East Branch trip included this theme.

In 1879, another conservationist, Theodore Roosevelt, a young man at the time, entered the country of the East Branch. He crossed the river on a trip to climb Katahdin, as so many did in the latter half of the 1800s and first years of the 1900s. As president of the United States, he would join the ranks of the country's greatest conservationists and would become one of the nation's strongest supporters of the national park system. In 1906, he signed the Antiquities Act which allowed presidents to designate national monuments, of which some, such as the Grand Canyon, became national parks.[19]

In 1861 scientists of the Maine Scientific Survey also visited the farm. The group was led by Dr. Ezekiel Holmes, naturalist, and Dr. Charles H. Hitchcock, geologist.[20] Hitchcock along with botanist George L. Goodale and entomologist Alpheus S. Packard stayed a few days at the farm doing scientific exploration while others were making observations elsewhere. Packard wrote that he and Hitchcock collected fossils in the vicinity of the dam on Matagamon Lake "which will enable him [Hitchcock] to determine the age of the rocks about here that have puzzled geologists a good deal."[21] The trip resulted in the first geologic map of northern Maine, and Hitchcock, in Rankin's words, "obtained a commendable grasp upon the stratigraphy along the East Branch and Matagamon Lake," and used fossils to date some outcrops in the region.[22]

Rankin also reported in his dissertation that "in 1905, O. O. Nylander made a number of fossil collections from sandstone along the East Branch above Haskell Deadwater and along Matagamon, Webster, and Telos Lakes." These were studied by the geologist J. M. Clarke in 1909 who concluded they were Silurian in age.[23] Thirty years later Nylander would be mentoring a young Bill Forbes in the science of paleontology.

Meanwhile, through the 1800s, the unrestricted heavy logging and "cut and move on" practices often left the ground torn up and covered with piles of slash. Inevitably in times of drought, human carelessness and lightning strikes led to forest fires. Three great fires in the area burned several hundred square miles in 1884, 1903, and 1916, accounting for the large number of bare mountain tops that Rankin encountered, which aided his ability to observe the rocks and geologic features that would have otherwise been beneath the ground cover.[24]

When the Great Northern Paper Company was chartered in 1899, the country around Katahdin took on an industrial aura. Development of the area proceeded rapidly with settlements, roads, dams, and forest harvesting. By the early 1900s, much of the area had been cut more than once and the ownership of the land was reduced to a small number of individuals and corporations. Roads had opened many once remote places, and with greater numbers of visitors, the mystique of the north woods began to fade. With millions of acres of forest land owned by mill owners who needed to have a steady supply of wood, the concepts of conservation and sustainability took on greater economic meaning.

By 1955 when Rankin began his PhD work, much of the area, however, still retained a wild character that would become even more so because of the dream of one man. That man was Governor Percival Baxter who in 1920 had also crossed the East Branch of the Penobscot on his way to climb Katahdin, and he, too, like Teddy Roosevelt who had preceded him by some forty years, would become known for his conservation ethic. That very year of Rankin's arrival, Governor Baxter completed his acquisition of land in the northern part of the park, some of which was in Rankin's study area, and in the next ten years, he would enlarge his land holdings to more than 200,000 acres. According to Baxter's wishes, most of the land would be permanently protected as "forever wild."

The 1955 addition of land included the historic Trout Brook Farm where Rankin and his new wife would stay for some of his mapping and research work. That acquisition also included the park's Scientific Management Area of nearly 39,000 acres for multiple-use management, including sustainable logging. This area lacked "forever wild" protection due to concessions Baxter was forced to make in order to gain the necessary support for its acquisition and acceptance by the State as part of his park. However, as we shall see, the Scientific Management designation, one day in the future, would help advance scientific research started by the fossil discoveries of Rankin and others.

The conservation story in the East Branch country doesn't stop there, for the efforts of both President Roosevelt and Governor Baxter influenced the dreams of another of Maine's great philanthropic conservationists, Roxanne Quimby. Through her purchase of thousands of acres around the East Branch and the eastern edge of Rankin's study area and through her subsequent efforts and those of her son, Lucas St. Clair, she was able to donate 87,500 acres to the federal government, which, in 2015 under Theodore Roosevelt's Antiquities Act, President Barack Obama designated as the Katahdin Woods and Waters National Monument.

This change in land ownership occurred sixty years after Rankin discovered the 400-million-year-old fossils in South Branch Ponds Brook—precursors of the forests he entered that day in 1955. Strangely, the plant remains were relatively undisturbed in a rock formation riding on a plate that had moved thousands of miles from south of the equator to well into the northern hemisphere and had endured the nearby history of unimaginable catastrophic forces of volcanism and land breakup. Yet, the evidence of plants once living at the edge of a tropical sea on a barren landscape barely hinting of the green it would be colored one day—that evidence had survived.

5

MAPPING A WILDLAND

In the summer of 1955, Doug Rankin laid aside his studies in the organized world of Harvard University and dove into the snarled tangle of dark woods in northern Maine. Here he would penetrate the roots of time to the first glimmers of the forest he entered, but without expectation nor forewarning, he would come upon what future scientists of the earth would call "one of the richest early land-plant assemblages in the world."[1] And well within a decade and a half later, others would follow his notes and maps, discovering the ancient ancestor of the trees hanging over the streams they waded—an ancestor that was only one of a kind, its evidence lying hidden in a curious formation. But all of this was unforeseen in 1955.

Rankin's principle objective was to commence the unraveling of the geological character of his study area and prepare a geologic map—essentially from scratch with the exception of a prepared mind. His map would be a bedrock map representing in different colors the kinds of rocks he found indicating their grouping into formations and showing the ages of rocks, their distribution and extent, and their relationships to one another. He would also make cross sections of rock formations showing their depths into the earth with estimations of how high they might have been before being eroded away. His task was also to imagine the history of the land and its formation, explain the processes that produced what he saw, and study evidence of early life fossilized in the rocks, which would help him determine the age of the rocks and kind of landscapes that supported life forms.

A geologic map, such as the one Rankin was going to do, would supply a piece of the puzzle of how his region of the earth developed over geologic time. Since a large part of his area was located in a wilderness park with many visitors, it would help identify the areas of geological and paleontological interest and study; point to appropriate localities for subsurface activity, such as road maintenance or locating groundwater resources; and contribute to understanding the relationship of bedrock composition and weathering to soil chemistry and plant ecology.

The basic information for Rankin's map could only be found through a field study. As such, his fundamental task was to find outcroppings of rocks. However, in an area as large as he was assigned to study, with its rugged terrain and difficult accessibility, he could spend a lifetime studying its geological details. And his task would not be made easier by the relatively short summer seasons in northern Maine, primitive living conditions, erratic weather, hordes of black flies and mosquitos, steep mountains and ridges, dense woods, thick swamps, and unstable bogs. But in the end, he knew that the map and accompanying information would rely on his best judgement and interpretation of the information that he could reasonably obtain given time, cost, and other constraints. And from his past experience in the park on the AMC trail crews, it would not have escaped him that he had a unique opportunity to enjoy the beauty, solitude, and quiet of Maine's largest wilderness area.

An indispensable aid in the field is the topographical map, but Rankin discovered when he prepared for his study that there was a lack of good contour maps of the Telos Lake and Traveler Mountain quadrangles, major areas he was to study. Only preliminary editions of the topographical maps were available and these had 50-foot contour intervals. Maps with these large contour distances could not provide good detail for areas with little variation in terrain. It wasn't until 1960, near the end of his dissertation work, that the more desirable 20-foot contour maps became available. To overcome this problem, Rankin used a variety of maps, aerial photographs, and an aneroid barometer to plot field data.[2]

An aneroid barometer measures change in atmospheric pressure which, in turn, can reflect a change in altitude. Therefore, such an instrument can be used to determine the elevations of outcrops. But atmospheric pressure is also susceptible to many factors, such as temperature and movements of high and low pressure systems, and so to make necessary corrections in the readings, the barometers need to be checked frequently at known elevations and with the time also recorded.

Rankin's preparation also required that he develop a system to record the location of his geological findings. He decided to locate each finding on one of his quadrangle maps. Each map would be divided into nine sections, and every important geological specimen or outcrop would be located in a section by marking a point measured to an accuracy of a tenth of an inch or a hundredth of an inch.

Thus, having made the necessary preparations and armed with the equipment he needed Rankin headed into the field. His initial primary focus was on outcrops accessible along the limited number of roads, the more prevalent hiking trails, wadeable brooks and streams, and the shores of lakes and ponds, which he could reach in many cases by canoe or boat. At each significant outcrop he came across, he measured the dip and strike and identified the rocks and minerals present. Sometimes he found himself "chasing formations," that is, looking across strike direction for outcrops of the same formation and other formations. At times, he made a grid map of areas between outcrops and used a "pace and compass" method as he bushwacked through the woods. At each significant location, he plotted it on a contour map according to his system. He also recorded his observations in a field notebook, which typically included sketches, measurements, and notes. Geologists today supplement sketches with digital photographs and use global positioning system instruments to record precise locations. They record their field observations on notebook computers or personal digital assistants (PDAs). Geologists are increasingly using software on notebook computers to draw lines in making field interpretations of the position of features, such as faults (cracks in rocks of the earth's crust where there may be movement on each side).

Rankin would spend nearly nine months of field work in the Katahdin-Traveler area during the summers of 1955 through 1958. When he published his study in 1961, he would say that those "summers of field work were made easier and more enjoyable by the assistance of . . . [his] father, Carl S. Rankin . . . and his wife, Mary Louise, who

Figure 5-1. Figure on felsite conglomerate on South Branch Ponds Brook 1950s. From: "Bedrock Geology of the Katahdin-Traveler Area, Maine," a thesis presented by Douglas Whiting Rankin to the Division of Geological Sciences for the degree of Doctor of Philosophy in the subject of Geology, Harvard University, Cambridge, Massachusetts, May, 1961. Published with permission: HU 90.8040.5 (Figure 81A). Harvard University Archives, and permission from the Douglas Whiting Rankin Family.

served ably as his field assistant for one full summer and parts of two others."[3] Years later, his wife would write that "the memory of those early years remains vivid. I witnessed and shared his love for the area during the first two summers of our marriage when I assisted in his field work. We feel particularly fortunate to have been able to spend some of that time in the now long gone gigantic farmhouse at Trout Brook Farm and at the also long gone work crew camp on the lake point east of Grand Lake Dam."[4]

The summer field work, however, was only part of Rankin's task. During winters, he conducted laboratory investigations of his rock collections. Over the course of his study, he examined 350 thin sections of rocks and minerals and undertook "a considerable amount of x-ray diffraction study."[5] A thin section is used to identify minerals by their optical properties. It is prepared in the laboratory by using a diamond saw to cut a thin slice of a rock or mineral, which is attached to a glass slide and ground flat and thin enough to become transparent or translucent. Then a second glass slide is added on top. Now the rock specimen, mounted between two glass slides, is examined under a microscope. For example, using this process, Rankin was able to report that sandstone specimens in the Trout Valley Formation were sparse in a feldspar mineral.[6]

Rankin also used X-ray diffraction to identify minerals in rocks. In the process, a beam of X-rays is passed through a rock powder produced by pulverizing a rock sample, in which the rock's different minerals each have a different crystal structure. The X-rays scatter and produce a distinctive pattern. Every mineral has a characteristic X-ray diffraction pattern that can be matched to an extensive database of patterns. Additionally, the percentage of the mineral in the rock can also be determined. Through this process, for example, he was able to confirm the identification of the mineral siderite, an iron mineral, in the rocks of the Trout Valley Formation.[7]

For six years Rankin worked to throw a geological net of understanding over his 300-square-mile area of exploration, and when he was done, he knew the geological makeup of the area—the layers upon layers of sedimentary rocks, the locations of once molten igneous rocks, and where

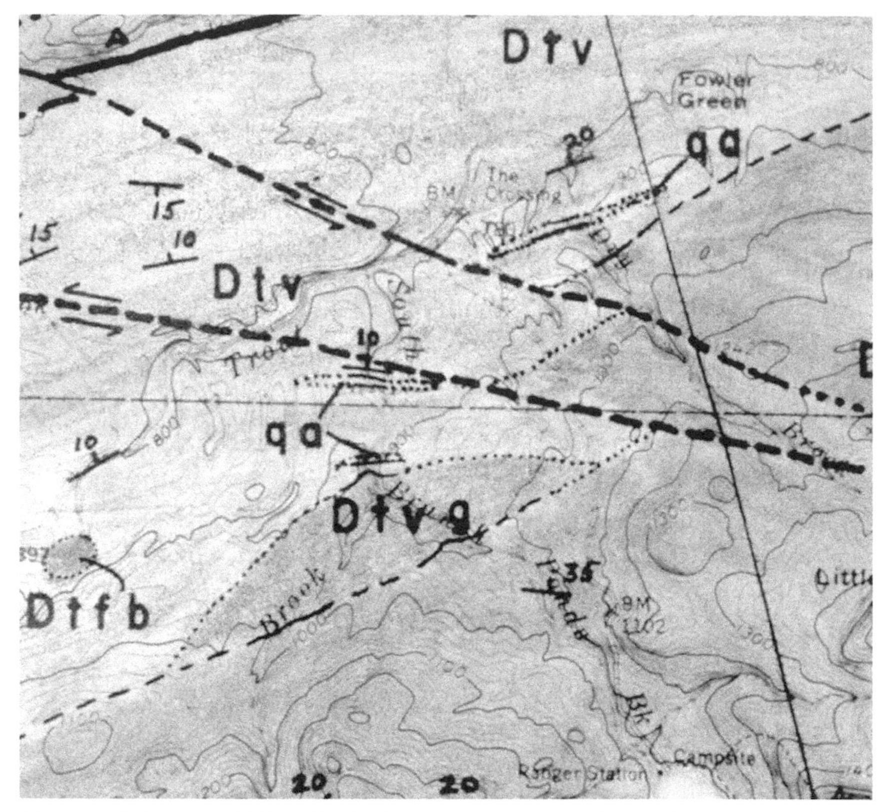

Figure 5-2. Douglas Whiting Rankin's geologic map, detail showing part of the Trout Valley Formation at confluence of Trout Brook and South Branch Ponds Brook. From: "Bedrock Geology of the Katahdin-Traveler Area, Maine," a thesis presented by Douglas Whiting Rankin to the Division of Geological Sciences for the degree of Doctor of Philosophy in the subject of Geology, Harvard University, Cambridge, Massachusetts, May, 1961. Published with permission: HU 90.8040.5. Harvard University Archives, and permission from the Douglas Whiting Rankin Family.

the rocks had been altered in structure and texture by enormous forces of heat and pressure to become metamorphic rocks.

In the end, Rankin had swept away some of the mystery of what had happened through the past hundreds of millions of years to create this place he had come to love. And through the years following his thesis work to his death in 2015, sixty years later, he continued to add to his knowledge through further studies and analysis. From him and the work of others, we now know that in the Ordovician Period, which began 485 mya, Rankin's study area was not part of Maine and the continent of North America; instead it was part of another continental crust in a large ocean which separated them.[8] As that ocean gradually closed because of continental drift, a portion of the earth's crust descended into the interior of the earth (known as subduction). At a sufficient depth, the crustal material melted and rose to the surface forming an arc of volcanic islands. The continued closure of the ocean resulted in collisions between island arcs and continents. These collisions, called orogenies, caused the eruption of volcanoes, deformation of rocks, and the uplifting of mountains.

It was in the middle of the Late Silurian Period, more than 420 mya, that an island arc of volcanic and plutonic rocks was closing in on the North American continent where Maine is now located. The name given to this arc or microcontinent, as it is sometimes called, is Avalonia and it collided with the continent at the end of the Silurian Period. The collision is known as the Acadian Orogeny, and it was a significant tectonic event of that time. In the words of one researcher and writer, it "was responsible for the formation of volcanoes, the closure of an open seaway . . . the genesis of the Appalachian Mountains . . . massive sedimentation events, and considerable underground magmatic activity."[9]

The tectonic front of the Acadian Orogeny's massive land deformation event rippled across what is now Maine in a north-by-north-west direction in enormous waves of earth movement—uplifting, folding, and contorting the land's surface. Ahead of the tectonic front the earth's crust was downwarped, forming basins of marine waters still connected to the ocean environment in which sediments from the erosion of highlands were collected.

The reconstruction of the timeline and geologic events related to the migration of the deformational front was a complicated scientific undertaking, as reported in 2000.[10] The researchers used data from the radiometric dating of rock samples by measuring radioactive elements and the decay of their radioactive constituents, for example, the decay of uranium to lead. The scientists also based their timeline reconstruction on strata containing the fossilized teeth of conodonts, eel-like creatures up to 40 cm long. The teeth, preserved as microfossils, were up to 5 mm long.

Scientists also used microfossils of an entirely different kind: palymorphs, which included fossilized spores, pollen grains, and marine microplankton, to date the rocks.

Based on the dating information the researchers determined that the tectonic front of the Acadian Orogeny migrated from the Bangor area to the area of Baxter State Park between about 423 and 407 mya. In the park location, a large body of magma was formed deep below the earth's surface. A violent volcanic eruption and great ash flows resulted in an enormous mass of the Traveler Rhyolite eight by twelve miles in area and perhaps two miles thick and containing eighty cubic miles in volume. It is hypothesized that a large volcanic depression, a caldera, formed in the vicinity of Traveler Mountain. The magma in the chamber beneath

Figure 5-3. After receiving his PhD in geology in1961 from Harvard University, Rankin spent the rest of his life exploring, interpreting, and mapping the geology of the earth enlightening both professionals and the lay public about the world around them. Photo of Douglas Whiting Rankin on Mt. Evans in Colorado in 2013. Photo courtesy of Yvette Kuiper.

the rhyolite crystallized to become the Katahdin Granite that forms Katahdin mountain which so many people climb today. In the words of Rankin and co-author Dabney Caldwell, in their booklet *A Guide to the Geology of Baxter State Park and Katahdin*, "Baxter State Park must have been an exciting place for a time in the Early Devonian."[11] According to Rankin, the erosion of the Traveler caldera began immediately after it formed and the sediments washing down its slope produced, through time, an unusual rock formation—the Trout Valley Formation which he had discovered at the beginning of his study in 1955.

In 1961 Rankin completed his thesis, received a well-earned PhD, and took a position at Vanderbilt University. The next year he joined the United States Geological Service and would spend the rest of his life exploring, interpreting, and mapping the geology of the earth, receiving many honors for his work and wide respect and admiration "for his judgment, dedication, good humor, and professionalism."[12] But throughout his life that rock formation, the Trout Valley Formation, that Rankin had discovered on his hike down South Branch Ponds Brook would draw him back to his beloved Baxter State Park time and time again.

6

A CURIOUS FORMATION

THE ANNOUNCEMENT CAME IN 1962, IN THE FORM OF A SCIENTIFIC paper, and it would draw some of the nation's leading paleontologists to the Traveler Mountain area of northern Maine. The paper had its roots in an event that took place in August of 1959, while Douglas Rankin was still working on his thesis study of the bedrock geology of the Katahdin-Traveler area. It was then that he had led a field trip in the area of the Traveler Mountain quadrangle for other scientists. The focus of the trip was on the curious formation Rankin had discovered in 1955, a large, mappable body of rock which appeared strangely unaffected by the volcanic activity, upheaval, and associated heat and pressure that had occurred nearby. But that was not the only thing that intrigued Rankin: it was the presence of fossils in the relatively undisturbed sedimentary layers of shale and siltstone in the formation.

William Forbes was included in the trip, undoubtedly because of his broad knowledge of fossils in the area. But the person who would play the largest role with Rankin was Professor Erling Dorf of Princeton University who would become Rankin's co-author on the 1962 paper.[1] Since 1930 Dorf had been a professor in the Department of Geology specializing in paleobotany and stratigraphy and had developed a highly respected reputation in these fields. By 1959 he had been recognized for furthering our understanding of the Pliocene paleobotany of California, of the ecology of tertiary forests of western North America, and of the early Devonian plants of Newfoundland. He had also studied changes in ancient climates.[2]

Figure 6-1. William Forbes was invited in 1959 by Douglas Rankin to join him and Erling Dorf on a field trip to study a curious formation Rankin had discovered on his 1955 trip down South Branch Ponds Brook. Photographer unknown.

It was in 1955, four years before his northern Maine field trip with Rankin, that Dorf had been one of the faculty tapped by Princeton to develop and present a lecture on an NBC educational television series called *Princeton '55*. Dorf's lecture was titled "Climates of the Past." From the video, one can see that Dorf, then age fifty, still retained the physique of the swimmer he once was as captain of the swim team and diving champion at the University of Chicago—lean and trim with easy, fluid movement. The video also allows us to see him as a teacher who could present a well-organized and interesting story of how scientists study fossils to reveal ancient climates.[3]

Dorf's teaching style and ability would be recognized in 1963 when he received an award from the National Association of Geology Teachers for distinguished teaching and in 1967 when he received another from the Association for "notable contributions to layman-directed or pre-college teaching and writing."[4] In 1964 his skills in communicating to a lay audience also became widely known with the publication of his popular account of the petrified forests of Yellowstone National Park, which was revised in 1981 before his death in 1984.[5]

Figure 6-2. Erling Dorf, paleontologist, who studied the Trout Valley Formation with Douglas Rankin and with whom he published a definitive paper about the formation and its plant fossils. Photographer: Orren Jack Turner. From: Erling Dorf; Historical Photograph Collection: Individuals Series, AC067, Princeton University Archives, Department of Special Collections, Princeton University Library. Published by permission.

Erling Dorf loved teaching and working in the field, so it was that in 1959 he accompanied Rankin on a trip to the locations in Trout Valley where Rankin had discovered, during his summers of field work from 1955 to 1957, "well-preserved specimens of primitive psilophytes . . . rare in North America."[6] The plants of the genus, *Psilophyton*, are extinct vascular plants, most of which have been found in rocks around 400 million years old (Emsian age—408 to 393 million years ago). Erling Dorf was very familiar with psilophytes, for in 1934 he had discovered the first verified occurrence in the United States of *Psilophyton princeps*—a species also found later in the Trout Valley Formation. Dorf's specimens came from the Lower Devonian rocks of Beartooth Butte in Wyoming. [7] But the first discovery in the world of this strange, simple plant had been made three-quarters of a century earlier in 1859 by Sir John William Dawson. Dawson's specimens came from the Gaspé Peninsula, Quebec, Canada—one of the first fossil plants ever discovered of Devonian age.

It was, however, on Dorf's 1959 trip with Rankin, a hundred years after Dawson's initial discovery, that the two successfully located good psilophyte specimens in a plant-bearing zone. The specimens were in an outcrop of black shale in a section of "soft crumbly sandy shale along . . . South Branch Ponds Brook."[8] Two years earlier, Rankin had discovered good specimens on "float," that is, loose pieces of rock not connected to an outcrop.[19] These were transported from their original location by water. But in 1959, the fossils discovered by Dorf and Rankin were in actual bedrock, found "in place" or *in situ*, a Latin phrase meaning "in position."

Fossil plants found in their original position in the bedrock—*in situ*—provide paleontologists the opportunity to gather much more information about the plants' ancient environment. This information might be about associated plants or animals living at the same time and place. The plants or animals found in rock strata *in situ* may provide information about the age of the strata—the geological "dating" of the rocks. In addition, scientists might be able to ascertain the relative abundance of the plants' patterns of growth, the environment they lived in, and depositional setting. Additionally, fossils *in situ* are more likely to be intact and are sometimes easier to spot in the rocks.

In April of 1961, after Dorf had studied the material collected on their 1959 trip, he wrote to Rankin that he had confirmed the identification of *Psilophyton princeps* var. *ornatum* Dawson along with three other types of fossil plants. Thus, in their 1962 paper, they reported the second verified occurrence of this fossil plant in the United States.[10]

Psilophyton princeps var. *ornatum* was the most significant fossil reported by Rankin and Dorf in their paper. They reported eighteen well-preserved remains of its spiny stems. Since Dawson's discovery of it in 1859, the plant has had an interesting and confusing history of classification. This demonstrates how the thinking of scientists evolves to accommodate new discoveries. For example, Dawson gave his 1859 plant the genus name, *Psilophyton,* because he thought it resembled the modern whiskfern, *Pilotum*. However, there is just a superficial similarity in appearance between the two but no evolutionary connection—this is based on our more recent understanding of the highly modified fern, *Psilotum*. New information requires new scientific ideas. Dawson gave his fossil the species name, *Psilophyton princeps*. Since *P. princeps* was the first-described species, it is designated as the "type" species—the example for any new and similar species.

The confusion began with Dawson's first description of *Psilophyton princeps* from the Gaspé Peninsula of Quebec, Canada, in 1859. Dawson's description was of fragmentary slightly spiny plant stems. A few years later after additional collections, he gave a more detailed description of the species with a drawing (called a "reconstruction"). This was to be the standard example in textbooks of an early Devonian plant for decades to come—a leafless plant with forking slightly spiny stems bearing sporangia at their tips. The confusion came in 1871 when, with additional new collections of fragmentary stems, Dawson described a "more spiny" plant which he called *Psilophyton princeps* var. *ornatum*—close enough to the original species but different enough to be called a "variety." It turns out, almost a hundred years later, that *Psilophyton princeps* var. *ornatum* is not a "variety" at all but a completely different plant species. This was the work of the paleobotanist Francis Hueber of the Smithsonian Institution. As mentioned above, the original species of Dawson had sporangia at the tips of its stems—reproductive structures are the key to identification and classification. Hueber reported in 1967 that the fossils of the "variety" carried their sporangia attached to the sides of the stems, not at the tips. To commemorate the groundbreaking work of Sir John W. Dawson, Hueber wanted to name the new species after Dawson, i.e., *Dawsonia*. Unfortunately, the name was already taken up by a moss. So Hueber scrambled Dawson's name into "sawdonia"—hence *Sawdonia ornate*—keeping the variety name also.[11]

Confusing, yes, but all in the interest of trying to be accurate in sorting out the world.

Lisa Churchill-Dickson in her comprehensive and illuminating book, *Maine's Fossil Record: The Paleozoic*, summarized the significance of the fossils: "The Trout Valley Formation . . . contained some of the oldest plant remains in the world [at the time of Dorf and Rankin's 1962 paper]. Older remains have been found since that time, but the flora and fauna of the Trout Valley Formation still play a critical role in our understanding of early terrestrial communities, their diversity, and structure."[12]

It was soon after Rankin had taken a position at Vanderbilt University, teaching and inspiring a new generation of students to field work, that he and Dorf published their paper titled *Early Devonian Plants from*

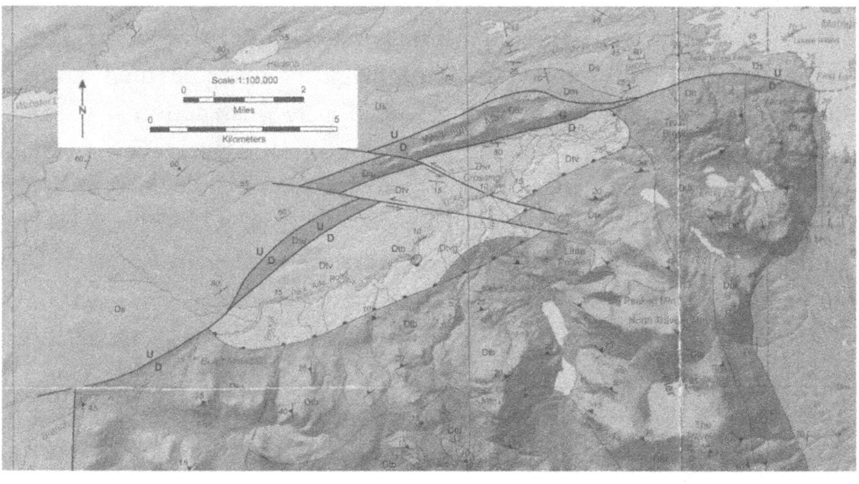

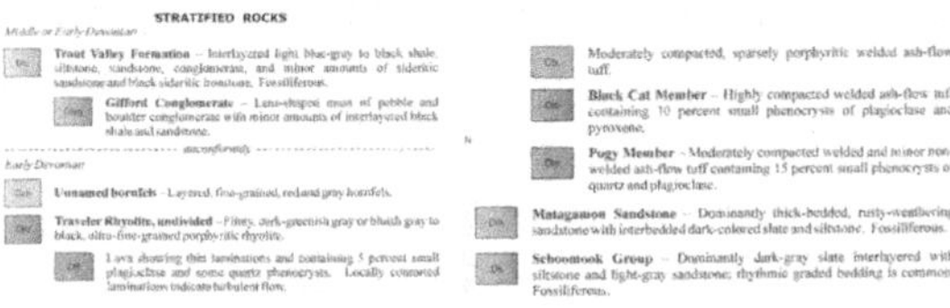

TROUT VALLEY FORMATION
Maine Geological Survey
Department of Conservation
2010

Figure 6-3. Geological map of the Trout Valley Formation, Bul. 43, Plate 1-Bedrock Geology, Douglas W. Rankin, Andrew Griscom, and Robert G. Marvinney, from *A Guide to the Geology of Baxter State Park and Katahdin*, Maine Geological Survey, Department of Conservation, Douglas W. Rankin and Dabney Caldwell. Published by permission of the Maine Geological Survey, Department of Conservation.

the Traveler Mountain Area, Maine. For the first time, the paper gave a published name to the Trout Valley formation, which, up to then, had received no official name. It also designated a type section for the formation and gave the locality for the fossil site "as the exposures along Trout Brook."[13]

Dorf and Rankin in 1962 described the rocks of the Trout Valley Formation as light-blue-gray to black and consisting of shale, siltstone, sandstone, and conglomerate sedimentary rocks. They indicated that many of the rocks are carbonaceous, rich in carbon, and most are slightly calcareous, that is, they contain calcium carbonate, the main component of such things as pearls, sea shells, egg shells, and lime used for agricultural purposes.[14]

Dorf and Rankin also noted that changes in the rock types of the formation are abrupt when viewed vertically. They measured and described a typical section of outcrop twenty-five feet high along Trout Brook. From the top of the section down five feet, they observed "sandy

Figure 6-4. Sheila K. Bennett explores an outcrop on Trout Brook showing a cross section of the Trout Valley Formation which Dorf and Rankin described in their 1962 paper about the formation. Photo by Dean B. Bennett.

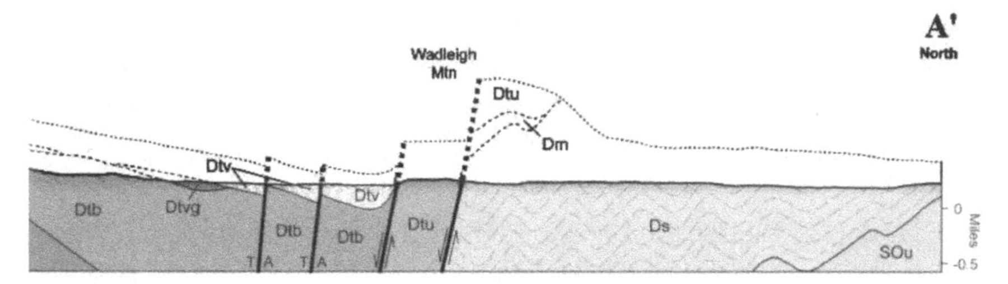

Figure 6-5. Intepretive cross section from the geological map of the Trout Valley Formation. Dotted lines show limits of missing rock layers and fossils they may have contained. From *The Geology of Baxter State Park and Katahdin*, Bulletin 43, Plate 1-Bedrock Geology, by Douglas W. Rankin, Andrew Griscom, and Robert G. Marvinney. Published by permission of the Maine Geological Survey, Department of Conservation.

black shale containing plant impressions," and continuing downward they found distinct beds of sandstone, shale, conglomerate, and siltstone. Below fourteen feet, they discovered a six-inch layer made up of "black carbonaceous shale containing abundant plant impressions." At the base of the formation along South Branch Ponds Brook and Gifford Brook, Rankin noted "a massive felsite conglomerate lentil (Gifford Conglomerate), about 300 feet thick."[15]

While Rankin could determine how deep the Trout Valley Formation was as it rested on the rhyolite bedrock below, it was more difficult to determine how high it might have extended because the layers of rock above were missing. The Trout Valley Formation with its fossils had developed during Devonian time, which ended about 360 million years ago. Since that time all of Maine's bedrock that may have contained fossils, all of its record of life the rocks may have had, disappeared. No fossils of dinosaurs or early mammals have been found because there are no rocks from younger sedimentary deposits of those times. What happened?

Lisa Churchill-Dickson, working for the Maine Geological Survey, describes three major causes for the absence of formations with fossils younger than Devonian times. First, the Acadian Orogeny, a mountain-building event in the late Devonian, replaced the inland seas that once covered areas of Maine, leaving the state predominantly as dry land.

Plants and animals were more likely to be preserved and fossilized in the underwater conditions by marine sedimentation than they were by the depositional processes that occurred on the land surface. Second, mountain-building events since the Devonian have caused deeper levels of fossil-containing strata to be uplifted and worn away by erosion in the last several hundred million years. Third, during the Pleistocene epoch, from about 2 million to 10,000 years ago, ice sheets up to two miles or more thick passed over Maine. Their tremendous weight and movement broke up the underlying layers of rock along with any fossilized evidence of plants and animals.[16]

It is easy to see why, with the rarity of Devonian fossil sites and the unusual opportunity to find Devonian fossils presented by the Trout Valley Formation, Dorf and Rankin were motivated to follow the scientific tradition of reporting their findings in a paper for publication in a scientific journal. This formality, begun in Britain and France in 1865, is a rigorous part of the scientific process of describing research results meeting high standards of reporting so they may be judged by peers. Research articles allow scientists to report new discoveries, to be kept up to date, to learn about prior research encouraging further studies, and to have their discoveries become part of the scientific record.[17]

Thus, soon after their study of the Trout Valley Formation and its plant fossils, Dorf and Rankin published their findings announcing "the first occurrence in New England of such well-defined members of the widespread *Psilophyton* Flora of the Early Devonian."[18] And it wasn't long before paleobotany took on a new interest and urgency in northern Maine.

7

THE FOSSIL HUNTERS

When that government jeep pulled up beside William Forbes at the road cut in 1964, he already knew two of the three occupants: Ely Mencher, MIT geologist, and James Schopf, U. S. Geological Survey paleontologist. But when he was introduced to the third scientist, Henry Andrews, a paleobotanist from the University of Connecticut, it was a pivotal moment in this story, for it ultimately led to the discovery of the only known locality in the world at the time of one of the tallest plants on the landscape 390 million years ago, now known as the Maine State Fossil.

Henry Andrews was fifty-four years old, about twenty years Forbes' senior, when the two met. Andrews was already well on his way to becoming known as one of the twentieth century's outstanding pioneers in paleobotany. Born in Melrose, Massachusetts, in 1910, his father a lawyer, and mother a housewife, Andrews grew up in a happy, comfortable home with parents who supported his interest in natural history, plant collecting, woodworking, and hiking in the White Mountains.[1] Andrews led a busy life in his Melrose school days with his friends and sports mates. His father is reported to have recalled that Henry "was always the leader, and without saying a word."[2]

After high school in Melrose, he completed a college preparatory year at the New Hampton School in New Hampshire, entered Northeastern University, and then transferred to MIT to major in food technology. This choice of a field of study would appear to have little relationship to the career he would forge in paleontology, except for two things: it

Figure 7-1. Henry N. Andrews, Jr. 1971. From the collection of Andrew E. Kasper, Jr.

provided "a very good biological-chemical, etc., education,"[3] he said years later, and it led to his discovery of what he wanted to do for his life's work when his major advisor allowed him to substitute a paleontology course in place of a program requirement. The course was taught by Professor

Harvey Shimer, a recognized authority on fossils. Andrews later wrote: "This was the turning point in my career. I loved fossils and, like many others, I loved Shimer."[4]

Andrews knew that if he wanted to study fossil plants he would need a solid background in living plants, so he spent a year of study at the University of Massachusetts. There, he met a professor from Washington University in St. Louis who offered him a teaching assistantship. He took the opportunity and arrived at the University in 1935 and began a master's degree. After completing the degree in botany, he spent a year at Cambridge University in England working with paleobotanist H. Hamilton Thomas. He then returned to St. Louis, where his interest in paleobotany had been whetted by the Pennsylvanian-age plant fossils he had found in the area. He began a PhD program in paleobotany, receiving a doctorate in 1939.[5] That year he married Elizabeth Claude Ham, a student he had met in an advanced botany class at Washington University.[6]

Wth his wife, Lib, as she was fondly called, at his side, Andrews' personality traits that endeared him to others appeared to have doubled. The two touched the lives of all who came to know them with their warm hospitality; their generous nature of sharing their knowledge, wisdom, and resources; and readiness to help those who came to visit—colleagues, students, friends, and family.[7] At the end of his life, Andrews was remembered for being a leader, for being a prized advocate for graduate students, for his unhurriedness—"making time for almost everything and everyone," for being a good listener, for a "quiet way of sharing interests," and for his generosity.[8]

Acts of generosity were not uncommon among the paleontologists the authors have known. For example, when one of the authors (Dean) met with William Forbes many years ago in his laboratory at the University of Maine at Presque Isle, Forbes gave him a small fossil of *Pertica quadrifaria*. It was special because it contained reproductive structures, sporangia, by which *Pertica* reproduced. Andrews, himself, had many experiences of generosity of that sort. For example, a year before he had received his PhD, he met another paleontologist who, in his words, "was always generous in sharing what he had and always anxious to place fossil-plant collections in the hands of those who he thought could make use of them."[9] His name was James Schopf and he had received a doctorate from the University of Illinois the year before Andrews had received his. The two developed a close friendship that would last throughout their lives.

In their early careers, both Andrews and Shopf carried out productive research in plant fossils from the Carboniferous Period. This was a time in earth history, lasting 60 million years, when there were great swamp forests distributed from the east coast to the midwest. These ancient forests produced the massive coal deposits we know today. Mixed with the softer coal were "coal balls"—spherical lumps of fossil plants preserved in a hard, mineral matrix ranging in size from several inches to over a foot in diameter. Coal balls were formed by the precipitation of minerals which cemented the sediment around a nucleus of material such as part of a fossil plant or animal. Though they are not coal, coal balls come from similar environments of deposition.

Andrews and Schopf became widely known paleobotanists and received numerous honors. Schopf became a professor at the University of Illinois and a researcher for the U. S. Geological Survey.[10] One day, Schopf would be instrumental in making a contact that would result in Andrews' leading role in the discovery of the Maine State Fossil.

In the 1940s Andrews joined the faculty at Washington University and became the paleobotanist at the Missouri Botanical Garden. He expanded his study of fossils in coal balls. In 1947 he published his first book, *Ancient Plants and the World They Lived In*. Through the 1940s and 1950s, he continued a busy workload of teaching, research, writing more than forty articles, and administration. In these years he took advantage of fellowships and other opportunities, and his career took him and Lib to Belgium, Sweden, Ethiopia, Russia, India, and many places in this country. His visits, lectures, correspondence, and activities in American and international organizations widened his reputation around the world.

In the 1960s Andrews shifted his focus from fossil plants from the Carboniferous to the earlier plants of the Devonian. During the summers of 1962 and 1963, he explored Ellesmere Island in the Northeast

Territories of Canada for Devonian fossil plants. Later he wrote: "I think one loves the Arctic or not at all." In this case, he loved it.[11]

In 1964 Andrews left his positions in St. Louis to become Head of the Botany Department at the University of Connecticut. It was in that summer that he joined Schopf and Mencher in their Maine Devonian diggings and met Forbes. It was Forbes who led Andrews to northern Baxter State Park and the fossils in the Trout Valley Formation. Andrews' work in Maine was to occupy him off and on until his retirement ten years later and would show up in his writings well beyond.

The connections among the geologists and paleontologists working in Maine were invaluable. The connections formed a wide, congenial, helpful network of social and professional information exchange by word-of-mouth, published papers, meetings, field trips, and a generous sharing of collections. So it was when Andrews came East in 1964, that James Schopf brought him into this active network of geologists and paleontologists. Scientists had only in recent years begun an ambitious geological mapping project and were increasingly discovering fossils in places where one report acknowledged that it had been "no secret that the forested areas of north-central Maine were not generally regarded as a likely place to look for fossil plants."[12] That year, 1964, Schopf had, in fact, just published a paper on a new collection of plant fossils in a formation of sandstone deposited after the earth-changing events of the Acadian Orogeny had limited the prevalence of fossils in the rocks of northern Maine. It should be noted that Schopf had referenced Rankin and Dorf's paper throughout his report. Of significant note, one of the new species described was named *Calamophyton forbesii* after Bill Forbes who contributed to the fossil collection.[13]

That same year, 1964, a large study by six scientists, including Forbes, was conducted with the aim of mapping the geology of the quadrangle to the northeast of Baxter State Park. In it Dorf and Rankin's paper was again referenced, bringing attention to the Trout Valley Formation and its fossil plants.[14]

In the fall of 1966, Dorf and Rankin's paper received more attention when Rankin led a field trip as part of the New England Intercollegiate Geological Conference (NEIGC) in Baxter State Park. These

Figure 7-2. Members of the New England Intercollegiate Geological Conference (NEIGC) relax during a field trip lunch break at South Branch Pond in Baxter State Park. The trip to the Trout Valley Formation, which included wading down South Branch Ponds Brook, was led by Douglas Rankin in 1966. From a slideshow in which Rankin was honored in 2015 by the Geological Society of America's (GSA) Northeastern Section. Photo by Peter Robinson.

conferences, traditional annual events, had begun in 1901 with the purpose of bringing together in the field geologists interested in recent geological work and learning about current problems of New England geology. A field trip guidebook for each conference was prepared with papers related to the theme and location of the conference along with maps and detailed descriptions of each field trip including step-by-step directions, mileages, and stops.

By 1966 the conference had grown to a point that the NEIGC and the conference planners were anticipating 250 or more attendees. The conference's focus that year was the Katahdin region and included an all-day, fifty-mile trip by vehicle with numerous stops. One stop involved a three-mile hike led by Rankin down South Branch Ponds Brook and Trout Brook to places where he had made his discoveries in 1955.[15]

Figure 7-3a. Douglas Rankin leading a Carolina Geological Society field trip in 1967, a year after his Maine trip down South Branch Ponds Brook. Photo Courtesy Duncan Heron.

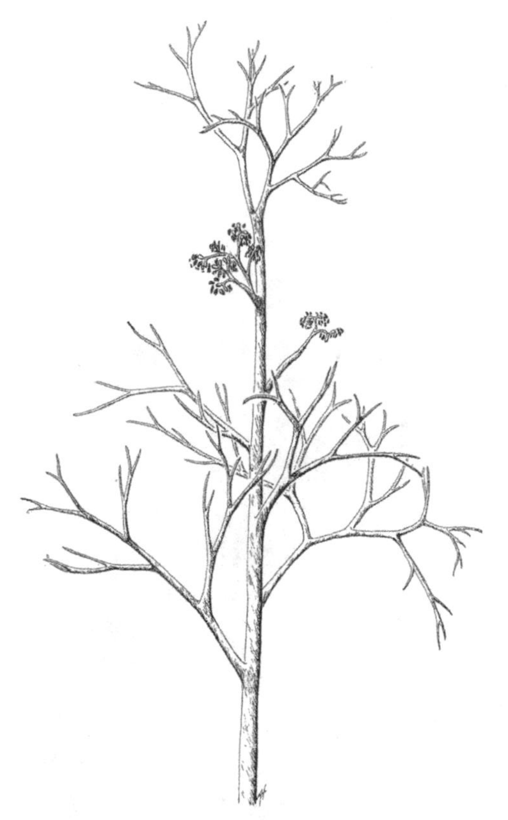

Figure 7-3b. *Psilophyton forbesii*, named after William H. Forbes. Illustration by Mary Hubbard, staff artist of Biological Sciences Group at the University of Connecticut, Storrs. From the collection of Andrew E. Kasper, Jr.

In the winter of 1968 Andrews, Kasper, and Mencher published a paper about plant fossils found along South Branch Ponds Brook in Baxter State Park. The prelude to this publication you will recall from the beginning chapter: Bill Forbes and the jeep with Mencher, Schopf, and Andrews. The paper described a new species of fossil plants based on a major collection made by Forbes and, appropriately, named after him: "*Psilophyton forbesii*, a New Plant from Northern Maine" . . . "in recognition of his knowledge of Maine Geology and his generosity in sharing this knowledge and his collections."[16]

The youngest of the three authors of the paper and a student at that time was Andrew Kasper. He was a graduate student of Henry Andrews at the University of Connecticut. Kasper was born and had his early schooling in Bridgeport, Connecticut; his high school education was in the Philadelphia area. He received his BA from Duquesne University in 1965 in philosophy and the classical languages, Greek and Latin. Having no resources and having been offered a small stipend to work in a laboratory, he entered graduate school at the University of Connecticut as an in-state student. The stipend and laboratory were provided by Henry

Andrews. Kasper wanted to be a mycologist—an area of biology dealing with fungi, and he was placed with a professor of mycology. However, Kasper realized that this was not a good fit, but working in Andrews' laboratory as a technician and benefiting from the previously noted generosity of Andrews, Kasper quickly changed direction and became Andrews' graduate assistant. Having a minimal background in science, graduate studies consisted of undergraduate and graduate courses with the constant encouragement of Andrews. He later admitted that he took the risk of accepting Kasper as a graduate student because with a classical background "he could read and write." Kasper's MS degree was completed in 1968 and his PhD in 1970. That year Kasper took a position at Rutgers University at the Newark, New Jersey, campus as an Assistant Professor of Botany teaching a variety of plant biology courses and paleobotany.[17]

In the summer of 1966 Andrews took his new graduate student to Maine for field research. Kasper clearly recalled his first field "test": going over a good-sized cliff with a rope around his waist trying to retrieve fossil plant fragments. He passed his test. Later he remarked to one of us

Figure 7-4. Andrew E. Kasper, Jr., late 1960s or early 1970, in Trout Brook area. From the collection of Andrew E. Kasper, Jr.

(Dean) that this all started with the 1962 Dorf and Rankin paper that had whetted interest in this area of northern Maine.[18]

In 1969 another paper was published based on the discovery of a new genus and species from one of Dorf and Rankin's localities in the Trout Valley Formation. The fossil plant was named *Kaulangiophyton akantha*—the genus name is Greek for "plant (-phyton) and sporangia (-angion) on the stem (Kaul-)." The second name, which is the specific epithet, is Greek for "spiny," i.e., the stems are spiny. This plant was discovered by Andrews and Kasper on August 1, 1967. At that moment, almost noon on a hot day along Trout Brook, Henry thinking about lunch, said "come on Kasper, find something." Reaching into a crevice in the rock, Kasper pulled out a fertile specimen of *Kaulangiophyton* and turned to Andrews saying "how about this." It was time for lunch.[19]

The discovery was published by a new graduate student in Andrews' laboratory, Patricia Gensel, with coauthors Kasper and Andrews. The study of *Kaulangiophyton akantha* was part of Gensel's master's thesis. The plant consisted of upright stems about a foot tall, about ½ inch in diameter with small sparsely scattered spines and large reniform (kidney-shaped) sporangia attached to the stems by a short stalk. Patricia Gensel would become a well-known paleobotanist.

Gensel was born in Buffalo, New York, in 1944. She graduated from Hope College in Holland, Michigan, earning a B.A. in 1966. She then worked for a year as a research assistant in London studying palynology (the study of fossil pollen and spores). In 1967 she began graduate school at the University of Connecticut with Henry Andrews, specializing in the study of Devonian and Early Carboniferous plants. She received her PhD in 1972, which was followed by three years of post-doctoral research with Andrews, primarily in Early Devonian plants in south-eastern Canada. In 1975, the year Andrews retired, she joined the faculty at the University of North Carolina, Chapel Hill, engaging in an active career of research, teaching, and writing that continues to this day. Her many accomplishments include the books *Plant Life in the Devonian* (1984) with Henry Andrews and *Plants Invade the Land* (2000) with Diane Edwards, numerous articles and research reports, and President of the Botanical Society of America for 2000–2001. She is also the

Figure 7-5. Patricia G. Gensel at an outcrop in the field. Photo courtesy of James E. Mickle.

namesake of the fossil plant *Genselia* Knaus, consisting of four species of early Carboniferous plants.[20]

In 1969 while Patricia Gensel was finishing up her report on the discovery of the new genus, *Kaulangiophyton*, Henry Andrews and Andrew Kasper decided it was time to write a broad review of the fossils found in the Trout Valley Formation. Review papers are common in science as they assess the progress of research in a field and give a historical perspective on work accomplished, as well as providing a direction for work to be done. The paper was published in 1970 by the Maine Geological Survey under the title *Plant Fossils of the Trout Valley Formation*.

This review paper is of particular interest in our story about the Maine State Fossil. Following a brief introduction to paleontology in Maine, the characteristics of early land vegetation, and a description of some of the fossils found along with their locations, the authors wrote: "In the summer of 1968 we located another very productive plant site on . . . Trout Brook." This is the most productive fossil site they had thus found. It was here that Andrews and Kasper found fossils that appeared to belong to a plant they had not seen before. The main stem was at least 1 cm in diameter and they speculated that the plant was at least several feet tall with regular-spaced branches and clusters of reproductive sporangia at the branch tips. They reported that the plant was distinctive from others they had found and they believed that it might be an unnamed *Trimerophyton*.[21] Andrews and Kasper did not know that they were on the verge of a discovery that would lead to legislative halls and recognition of their work by audiences far beyond the boundaries of Baxter State Park where they had been working.

8

THE DISCOVERY

By the summer of 1968, Andrew Kasper was in his third year of collecting plant fossils in the Trout Valley with Andrews, Forbes, and others. It was an exciting and rewarding time for the young doctoral student. The routine was by now fairly settled and efficient. Their home-base was the Shin Pond House, a distance of only twenty to twenty-five miles or so from their collecting site in northern Baxter State Park but, still, a time-consuming journey over a narrow, twisting, gravel road. The Shin Pond House, however, was a comfortable multi-roomed lodge offering breakfast, a prepared lunch, a good supper, a chance for a shower, and an opportunity to read for a while, before turning in to sleep in a comfortable bed.[1]

When they headed to their study site, the team's work dress included high-topped boots with rubber soles, dungarees, shirt, and a hat for protection against sun and flies. Fly-dope was a must. Their equipment included a backpack, geologist's hammer, a two-pound sledge, and chisels to break open rocks. At the study site, the fossil hunters would wade up and down the brooks and streams checking rock exposures on both sides where the swiftly flowing water had washed away the soil and vegetation. They also looked for fossils in the rocks at the bottoms of the streambeds where visibility permitted. They were constantly alert for signs of fossils, looking especially for lines on the rock surfaces, indicators of fossil plant stems. Finding a promising rock and knowing that plant fossils can cause a weakness in the rock, they would use a sledge and chisel to split it into two halves. Each half was called a part and counterpart which, if

Figure 8-1. Henry N. Andrews taking a break from fossil hunting at campsite/ Ranger Station at Lower South Branch Pond, Baxter State Park, late 1960s. From the collection of Andrew E. Kasper, Jr. Photo by Andrew E. Kasper, Jr.

successful, would reveal impressions of the fossil plant on each half. After working three to four hours at a site, there might be as many as 50 to 100 specimens lying around. The paleobotanists would look over what they had, pick out the best specimens, wrap them in newspaper, fill the packs, and, in Kasper's words, "slog out."[2] This is the daily routine that led Andrews and Kasper to the discovery of the Maine State Fossil.

The details of the discovery were recorded by Andrews in his field notes which were typed-up by Kasper. A copy was given to one of the authors (Dean) during his interview of Kasper on March 7, 1988:

> *July 17, 1968: Spent the afternoon checking some of Rankin's localities along Trout Brook. . . . Found plants at most of the outcrops. . . .*
>
> *The R3 locality proved by far the most exciting with abundant "Trimerophyton." . . . Called Trimerophyton because it was the only large-stemmed psilophyte known at that time. This is a low ledge only*

Figure 8-2. Henry N. Andrews and Andrew E. Kasper, Jr., wading in Trout Brook in search of fossil plants. From the collection of Andrew E. Kasper, Jr.

a foot or two above water level. Being excessively hot we stopped at about 3:30—continue next day."

July 18, 1968: [Andrews and Kasper; Andrews' field notes] "Spent the day at . . . locality. . . . Very abundant specimen [sic] *of Trimerophyton-like plant—strong monopodial axis with side branches—and also obtained numerous fertile specimens—sporangia in characteristically fan-shaped cluster at the tips of ultimate branchlets.*

This is an ideal locality in several ways—the plants occur in low ledges that are within a few inches of the water and only 2-3 feet above the brook—quite a few sq. yards of rock exposed. For the most part only the top 4-6 split well—over much of the area impressions are evident and we took only part of what is available.

With the exception of one very small scrap, showing delicate [sic] *branchlets, everything here appears to belong to one species."*[3]

Figure 8-3. Henry N. Andrews kept details of the discovery of the new fossil in his field notes. From the collection of Andrew E. Kasper, Jr.

In discussing the find in an interview, Kasper said that he remembers vividly being very excited about the discovery, exclaiming: "Well, here it is, my PhD project!" Henry, always calm, even got excited, too, he said. The plant was clearly different.[4] The fossils were imbedded in fine-grained shale and preserved as either compression fossils with carbonaceous film from the original plant or as impression fossils exhibiting an imprint or outline of the plant. The fossils commonly had a yellow-orange-brown appearance that contrasted nicely with the surrounding rock.

Getting the specimens to the vehicle was far from easy. While a map indicates that the brook runs right beside the road, up to its very edge

Figure 8-4a. Andrew E. Kasper, Jr., at 1st discovery site of the new fossil (the Maine State Fossil) at Trout Brook, July 1968. From the collection of Andrew E. Kasper, Jr.

in places, it is in fact forty to fifty feet below the road. The bank down to the brook is exceptionally steep, nearly vertical in some locations, and densely covered by trees and shrubs. The authors in attempting to follow where the scientists had gone had to tie off a strong rope at the top of the bank and hold on to it to get down—imagine trying to get up with a load of fossils. Andrews and Kasper had their work cut out for them backpacking heavy slabs of fossil specimens to their vehicle—several trips up and down were needed.

Having the fossil in hand in the laboratory is just the beginning of months and often years of study—in this case two years to be exact. The question of geological age was one of great importance to Kasper. He noted that Rankin relied on two types of evidence to date the Trout Valley Formation which would provide the age of the plant fossils. One type of evidence was the area's stratigraphy and the other, the fossils in the general region.

Addressing stratigraphy, Rankin found that the Trout Valley Formation overlies the igneous Traveler Rhyolite (also called felsite by Rankin),

Figure 8-4b. The rock slabs containing fossils were carried in a wicker basket weighing over 100 pounds up a steep slope behind the 1st fossil discovery site to the road above. Photo by Sheila K. Bennett.

so it must be younger than the rhyolite. In turn, the rhyolite overlies the Matagamon Sandstone Formation. Thus, the rhyolite, sandwiched between two formations, was intermediate in age. With this information, Rankin could work out the relative ages of the strata. By his own study of the geologic events of the area and by the related research of other scientists, Rankin gained an insight into the geological age of the formation. Since Rankin's work, there have been many published studies of the Katahdin region's geologic history. It is now determined that the Trout Valley Formation is lower Middle Devonian or Eifelian Age, 388–393 mya; the Matagamon Formation is the older Pragian Age, 408–411 mya; and in the middle is the Traveler Rhyolite of Emsian Age, 393–408 mya. All are in the Devonian Period, and as Rankin concluded in his dissertation, all were formed close to the same time. Now we know that these formations cover a span of 23 million years.

Figure 8-5. The new fossil in its rock matrix. Photo by Dean B. Bennett.

The second type of evidence Rankin considered in dating the Trout Valley Formation relied on the presence of plant fossils in the formation. He noted that, to be meaningful, the fossils must be fairly complete and their ages cover a relatively short span of time. But, although the fossils found in the formation were fragmentary and had lengthy time ranges, they were Lower Devonian in age. Today, they are determined to be lower Middle Devonian in age, 393–388 mya.

One of the first laboratory tasks Kasper faced was excavating the plant to determine its morphology, that is, its form and structure—what it looked like when it was alive—its parts and their shape, size, and other characteristics. This information would ultimately suggest the plant's name and its place in the evolution of plant life. Fortunately, Kasper had an abundance of fossil material to undertake the study.

To show the plant's morphology, a process called *degaging* was used to expose the fossil. Under a binocular microscope using a small hammer, needles, and a syringe to blow rock particles away, the plant fossil was uncovered. Henry Andrews was introduced to this process in 1958, ten years earlier, when he was in Belgium collaborating with Suzanne Leclercq, a paleobotanist known for her study of Devonian plants. There, his visit overlapped with Harlan P. Banks, an American paleobotanist and professor at Cornell University, who was working with Leclercq for a few months.[5] In a 1972 paper, Banks wrote a brief tribute to Leclercq and her degaging technique, writing: "She early recognized the obvious but frequently neglected truth that part and counterpart constitute a plant—*not* one of these alone. This led her to the equally obvious, but again usually neglected, truth that the three-dimensional character of a living plant lies buried in the matrix."[6]

This is what Kasper sought, the plant buried in the rock matrix. He was fortunate that when he split one of the slabs of rock, it split down the middle of the fossil leaving him with mirror images, like opposite pages in a book. It would take him at least a year before he would see what the rock was hiding. Because the rock matrix was so hard, a Dremel electric engraver with steel phonograph needles was used to remove thick layers of rock from around the plant. The needles would have to be sharpened on a whetstone regularly and two Dremel tools were used alternately because they heated up. To protect his ears from the noise, Kasper put cotton in them.[7]

Kasper would start from a small segment of a branch exposed at the surface of the rock and excavate in both directions hoping to connect with the main stem and branch tips. The key was to show all the parts connected, but because the plant was initially buried in the sediment with numerous fragments, he knew that parts in close proximity were not necessarily connected. It was slow work, taking long hours between taking courses and student teaching. It was delicate work and he had to be extremely careful not to destroy sporangia and other details. At times it could be very frustrating because occasionally the tool would slip and chip off a plant part. In uncovering the small sporangia, he would revert to using a tack hammer and sewing needle. But overall, he said that he loved the work; it was very relaxing, like listening to classical music. You forget about everything and all your cares in the world; when you get into this, this is all there is.[8]

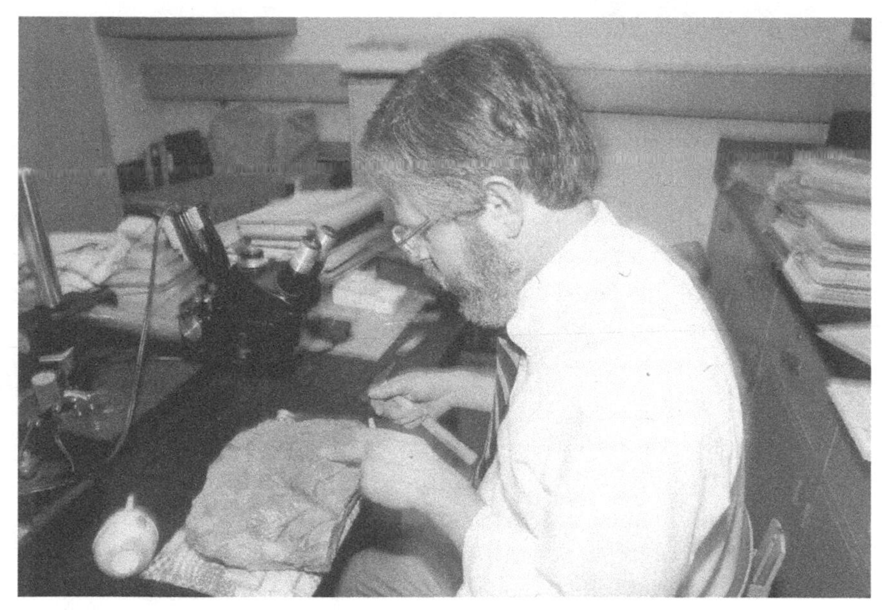

Figure 8-6. Andrew E. Kasper, Jr., demonstrating the degaging technique in his laboratory in 1988, twenty years after the discovery of the Maine State Fossil. The technique removes rock around the fossil to reveal some of its three-dimensionality. Photo by Dean B. Bennett.

The next step in Kasper's work was to produce a reconstruction, a life-like drawing of the plant, by a scientific illustrator—a task well beyond most scientists. The reconstruction was based on photographs, line drawings traced from the photographs, and accurate measurements of stem and branch width and sporangia size along with measurements of distances between the forks (dichotomies) of the branches. The measurements were carefully and constantly checked against the actual specimens.

Since the fossils were originally three-dimensional living plants, their branches, when pressed flat in preservation, covered one another. This required a special approach and technique. After the surface-exposed branches were degaged/excavated, they were photographed and then *destroyed* to expose the deeper branches in the rock matrix. This second deeper layer of branches was then excavated, photographed, and destroyed to expose the still deeper branches, and so on. The sequence of excavation, photographing, more excavation and photographing continued until one whole branch system was recorded on the *part specimen.* The same technique and sequence were done for the *counterpart specimen.* All photographs were then traced at the same scale onto semi-transparent tracing paper, constantly comparing with the actual part and counterpart. Up to four tracings from each part and counterpart were then placed in order on top of one another. The eight layers of semi-transparent tracing paper showed the 3-dimensional arrangement of the branch system true to scale. A final line-drawing was made of the combined tracings along with measurements.

The same technique, photographing part and counterpart of the main stem and branch locations, tracing the two photographs to scale, and placing the two tracings on top of one another, showed the plant's main stem arrangement. A final line drawing was made of the two tracings along with its measurements. The branch system (line-drawing) was then "attached" to its main stem (line drawing) along with the measurements. This stick-like drawing was given to the botanical illustrator at the University of Connecticut—Mary Hubbard. Kasper asked her if "she could make it look like a living plant," which she most certainly did.

Following are a few general comments Kasper made to us regarding the reconstruction using the well-known geology axiom: "the present is the key to the past." 1) This holds true for biology; 2) the the fossils were fragments—even though they were the largest specimens of a psilophyte known at that time; 3) the true height of the plant was estimated based on the stem diameter at about four- to six-feet tall; 4) the clusters of sporangia, although connected to the main stem, were numerous and distributed on a stem many feet tall in an unknown pattern; 5) the majority of the sporangial clusters were placed at the top of the plant presuming a better spore dispersal.[9]

Next, Kasper had to consider whether the plant was, indeed, a new species, consider the plant's classification and, if new, what to name it. After an extensive comparison with other Devonian fossil plants including those in museums in London and Montreal, Quebec, and since he was intimately familiar with the literature on Devonian plants at this stage of his studies, Kasper was confident that he had found a new plant. It was unique—a new species in a new genus. As regards classification, that was not a problem because of the plant's resemblance to the psilophytes: *Trimerophyton* and *Psilophyton*. It was quite clearly in the Subdivision Trimerophytina. The Trimerophytina belonged to the plant taxonomic system developed by Harlan Banks of Cornell in 1968—the very year of *Pertica*'s discovery. Now a new name had to be furnished for the new plant. Kasper told one of the authors that the fossil plant with its pronounced, central stem reminded him of a broomstick. The closest thing he found in a Latin dictionary was "pertica," which means a long pole or rod. In considering the second part of the species name (the specific epithet), he noted the branches and their regular pattern of attachment. In a bird's-eye view the branches were in four distinct vertical rows, hence the specific epithet, *quadrifaria*: "faria" meaning "bearing" or "carrying" and "quadri-" meaning "four." So the Latin name, *Pertica quadrifaria*, loosely speaking, means a "rod-like plant bearing four vertical rows of branches." Finally, after years of struggling with Caesar and Cicero as a student, Kasper could use his Latin skills in creating a new species name.[10]

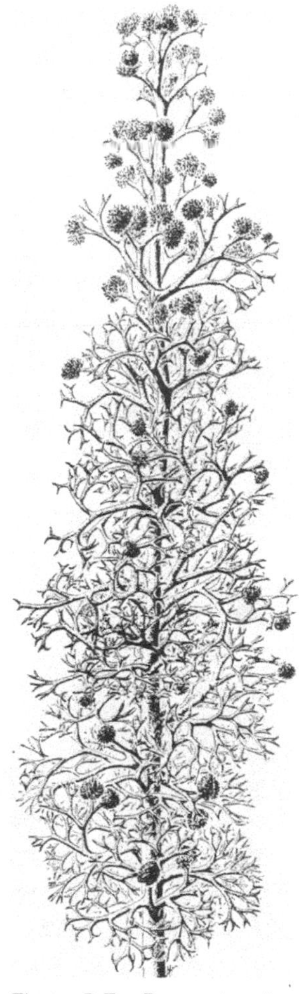

Figure 8-7a. Reconstruction drawing of *Pertica quadrifaria* by Mary Hubbard, staff artist of the Biological Sciences Group at the University of Connecticut, based on Kasper's study.

The discovery of *Pertica quadrifaria* was very important at the time. It allowed one of the most complete descriptions of an early land plant and permitted a relatively accurate reconstruction of its form or morphology. Kasper was able to describe the wide straight main stem, the arrangement of branches into four vertical rows, the equal forking (dichotomy) of both sterile and fertile branches. Finally, and importantly, there was the fact that the sporangia were numerous, in clusters at the ends of the fertile branches and in direct connection with the main stem. Zimmermann (1959) in his Telome Theory for the evolution of plants proposed that one possible origin of leaves was from the forking

Figure 8-7b. William H. Forbes assisted in the discovery of the second locality of *Pertica quadrifaria*. Photo from a tribute to William (Bill) Forbes slide program given in 2016 at an International Appalachian Trail meeting. No information on photographer given.

branch tips of Devonian plants which became "flattened" (forking in one plane rather than 3-dimensional) and "webbed" (filled with plant tissue in between the branchlets) to become leaves. The reconstruction of *Pertica quadrifaria* fits this scenario in plant evolution of leaves.[11]

In 1971 Kasper was a second-year assistant professor at Rutgers University in New Jersey and a research associate of Henry Andrews in the summer. Both paleobotanists were continuing their research in Maine and explorations for fossil plants in the Trout Brook area. That summer they were joined in the field by William Forbes. Working together as usual in the field, the three found a second and remarkable locality for *Pertica quadrifaria.*

As Kasper recalled, it was about noon, and Andrews was back at the van resting or eating lunch and Forbes and Kasper were slipping and sliding up Trout Brook outcrops as they stumbled along. Forbes suggested that they head back for lunch since both were tired and hungry. Kasper agreed, but asked for one quick look around the bend in the stream. Shortly thereafter, around the bend seeing vertical striations on the surface of a large block in the stream under water, Kasper gave a shout. A line of chisel blows at one end severed the rock into two perfect halves with three large stems staring the two scientists in the face. The fresh rock surface was a dark blue and the three stems and branches were maroon—each half a spectacular museum-quality specimen about 18 inches tall and 25 inches wide. The two excited paleontologists headed back to the van—thoughts of hunger had disappeared as they informed Andrews of the find.

We return to Andrews' notes:

> *July 21, 1971, Wednesday: Drove in Bill's car to locality of yesterday. . . . Collected until about 11 A.M.—I remained and dug out a few more pieces of* "P. princeps," *etc. and Andy and Bill continued exploration of Trout Brook Upstream. They returned about 12:45 with reports of a great find. After lunch we drove to this new locality. . . . There is a cliff 20–30 feet high on . . . brook. We walked downstream . . . and found abundant blocks* <u>*in the stream*</u> *containing spectacular* Pertica *(most abundant). . . . We hauled out 3 loads (Andy and Bill)*

of big excellent blocks containing spectacular specimens of Pertica. *Bill found origin of these blocks . . . rather close to the water level. It is quite remarkable that such large blocks, containing such fine plants could be battered about in the stream and remain intact.*[12]

One last comment by Kasper about the event: the blocks containing the fossils were so large and heavy that he and Forbes had to rope and carry one block at a time slung from a tree which served as a pole between the two paleobotanists as they walked mid-stream slipping and sliding to reach the vehicle—this was done again and again until they had an excellent collection and were utterly exhausted. But as a result of their efforts that day, this second *Pertica* site was discovered soon enough to make it into Kasper and Andrew's 1972 paper about the fossil.[13]

When the paper was published it contained another small note, probably insignificant to most but of great importance to a member of their team and to the future of paleontology in Maine. That member of their team was William Forbes, and the note was in the acknowledgements

Figure 8-8. Site of 2nd discovery of *Pertica quadrifaria* on Trout Brook in July 1971. Photo by Dean B. Bennett.

section of the paper. It said: "The writers wish to acknowledge the valuable assistance in the field of Dr. William Forbes (University of Maine at Presque Isle)."[14] It was not unusual to see Forbes' name in the acknowledgements of related geological and paleontological papers about the northern Maine region, but this was different. Up to this time he had been referred to as "Mr." William H. Forbes of "Washburn, Maine," rather than as "Dr." William H. Forbes of the "University of Maine at Presque Isle." The change was the result of events two years earlier.

January 1970 found Forbes on a personal mission: he wanted a college teaching and research job. This prompted him on the 24th of that month to put into his typewriter a sheet of his business stationery headed with the words "The Village Shoppe department store." On it he typed a letter to his long-time friend and well-known fellow paleontologist Dr. Arthur Boucot, Professor of Geology, at Oregon State University:

> *Dear Art,*
>
> *I really enjoyed talking with you on the phone last night. I am enclosing photocopies of a couple of letters from Hall and Boone. This may be of some help. With this lack of a degree business on my part, it will take some fairly strong letters of support to overcome the administration. So far it looks favorable.*
>
> *Aroostook State College is a small school, 1300 students, that has recently been incorporated into the University of Maine. They are in the process of expanding the Science Department to include Earth Science as a separate course. . . . The primary mission of Aroostook State College as a four-year school is for training elementary and high school teachers. I would also like the people there to be aware of the fact that it would be to their advantage for them to provide time and space for me to carry on research.*
>
> *With warmest personal regards,*
> *Bill*
> *Sent to Mr. Arthur Mraz [President of Aroostook State College]*[15]

Five days later President Mraz received a letter of recommendation from Dr. Boucot in which he said:

Mr. William Forbes, Washburn, Maine, has asked me to write to you concerning his qualifications as a potential teacher of your Earth Science course. Bill Forbes, in my opinion, is a very unique individual. The citizens of the Presque Isle region know him well and presumably think of him as a likeable, hardworking, friendly young salesman who could sell an icebox to an eskimo. What many of them may not realize, however, is that there is another Bill Forbes. This other Bill Forbes has never had the advantages of the Ph.D. level of education which his native intelligence and interest in hard work could have obtained for him. This other Bill Forbes through constant association over a period of almost twenty years with geologists of the U.S. Geological Survey and many universities has developed a competence and training above the level of many men who had the benefit of formal training. . . . His work on fossil conodonts, conducted as a labor of love, in his garage as well as in the basement of The Village Shoppe, is the most exemplary example of what a dedicated man can accomplish in science if he has the will power, native intelligence, and enthusiasm to carry it through. . . . I cannot think of another man better suited to your needs."[16]

On January 21, 1970, another letter arrived at the University of Maine at Presque Isle in support of Forbes' application for a teaching position. This time it was from Associate Professor Gary M. Boone in the Department of Geology, Syracuse University, who identified himself as "one of a large group of earth scientists in the United States and Canada who have followed Bill's rapid progress toward specialization and eminence in paleontology and regional stratigraphy in the northern Appalachian geologic framework. . . . His dedication to thoroughness and excellence in his work is renowned. . . . His infectious enthusiasm will bring to the College an immeasurable addition to the mystique of effective teaching."[17]

The next day, another letter arrived from Associate Professor of Geology Bradford A. Hall of the Department of Geological Sciences at the University of Maine. Hall wrote:

> *Bill Forbes is one of those rare people who reach a high level of competence in a field of science without the benefit of formal academic training.... The thing that had raised Bill well above the amateur status is his diligence and ability coupled with constant association with a good number of the very best geologists and paleontologists active in the field today. Bill has, in fact, received very good "formal" training informally, both in the field and in various laboratories. I doubt if anyone has a better knowledge of the total spectrum of Maine paleontology than Bill.... The University, the State, geology, and, I think, your students will benefit greatly if he is appointed to your staff.*[18]

Before that year of 1970 closed, Forbes was appointed a part-time lecturer at Aroostook State College of the University of Maine (eventually to become the University of Maine at Presque Isle). Within a year, he had advanced to a full-time geology instructor and had received an honorary Doctorate in Science from Ricker College. Fifteen years later, in 1985, he was promoted from Associate to full Professor of Geology. By then Forbes also held the title of Maine State Paleontologist, and in that capacity exercised an important role in bringing *Pertica* out from an obscure subject in a stack of scientific reports to a place of importance as a State of Maine symbol.

Figure 8-9. William H. Forbes as a professor at the University of Maine at Presque Isle. Photographer unnamed.

9

THE MAINE STATE FOSSIL

On a June morning in the spring of 1984, Clark T. Irwin, Jr., a reporter for the *Press Herald* in Portland, Maine, was having a cup of coffee while reading the *New York Times*.[1] He came across a story out of Albany, New York, where the Legislature had adopted as that state's fossil *Eurypterus remipes*, a giant extinct species of crab that "prowled the warm seas and lagoons of western New York when the area was submerged about 400 million years ago. The New York lawmakers rejected several other proposals for a state fossil, including one calling for the Legislature [itself] to win the designation."[2]

Irwin set his cup down and began to wonder: Did Maine have a state fossil? A call to state geologist Walter Anderson confirmed that Maine lacked a state fossil. Anderson suggested that "an ideal candidate rests in the compressed ancient sediments—shale, siltstone, sandstone—of the Trout Valley Formation in the northeast corner of Baxter State Park." Here, he said, have been found "some of the best examples of Devonian plant fossils in the world"—referring to the period of geological history from about 420 to 360 million years ago. One is a plant called *Pertica quadrifria*. It was first described in Maine and is, "without any question, one of the most significant finds in the northern Appalachians."[3]

Irwin then contacted University of Maine paleobotanist Professor George Jacobson who confirmed that "the Trout Valley Formation contains the mineralized remains of some of the earliest land plants in the world, and some of the best preserved." Jacobson also thought *Pertica* would be appropriate for Maine but suggested that the best person to make the nominating speech was *Pertica*'s co-discoverer, Professor William H. Forbes of the University of Maine at Presque Isle.[4]

A call to Forbes yielded pay-dirt for the reporter—and more. Forbes, Irwin learned, was a story by himself, "a fossil expert," in the words of Irwin, "as rare as his subject." He was a tenured professor of geology at the University of Maine at Presque Isle but most interestingly, he had no bachelor's degree, no master's degree and no doctoral degree, only an honorary doctorate from Ricker College—a small northern Maine college that had closed in 1978.

Irwin dug deeper and Forbes acknowledged that his lack of academic credentials had been a problem, but it seems he had resolved it by knowing more about his field than anyone else around. It was a love for rocks and fossils when he was a little kid that had dominated his life and led him to a career in geology and paleontology. He eventually had a number of Devonian fossil plants named after him.

Retired Colby College professor Donaldson Koons, a highly respected geologist, told Irwin that Bill Forbes was a "remarkable person." The U.S. Geological Survey "kind of forgot its rules" about credentials so it could hire him for field work. He was a researcher with "no equivalent in the Northeast." Professor Harold Pestana, a colleague of Koons, concurred in his assessment of Forbes, saying he was "a thorough professional" and "probably the single person most knowledgeable about the fossils of Maine."[5]

Forbes agreed that *Pertica quadrifaria* was the best candidate for Maine's official state fossil. He told Irwin that "*Pertica* is, or was, 'kind of exotic,' a leafless, spore-bearing, apparently rootless, curly-branched plant, up to 5 feet tall that grew in marshes. The plant's profusion of antler-like branches was 'a new strategy working toward a bona fide leaf.'"[6] Forbes shared other reasons for his choice with one of us in an interview in 1987. First, he said, it is unique to Maine: the type locality, where the fossil was first found, is in Maine and the species is only known from Maine. He believed that *Pertica quadrifaria* is especially fitted to be the state fossil because it is representative of the origin of modern vascular plants. The state tree, the white pine, is an example of a vascular plant with massive vascular tissue—wood. This tree and others in Maine's forests are important to the state's economy. Having a state fossil, Forbes said, gives another bit of evidence of the uniqueness of Maine and focuses public attention on the wonders of science and why science is important.[7]

Forbes was just the person to advocate for science. The "wonders of science" were at the core of his being and had been ever since he was a child. Like all engaged professionals, it was curiosity that drove him as it did the field of science. Science has taught us that we depend on plants for our survival. Knowing as much as we can about them—their evolution, their requirements for growth, their adaptations, their ecological relationships with the environment is fundamental to our stewardship of them. The study of how and why plants have changed from earliest geological times based on fossil evidence is also important to that understanding just as understanding past climate change is an important concern we have today.

Forbes was quick to recognize that a state fossil would raise public interest in and support for the study of ancient life and geology in general. He was, indeed, the person to choose the state fossil. Since 1950 he had been "conducting detailed field investigations to locate significant fossil locations in the state."[8] In 1977 at the invitation of the Maine State Planning Office's Critical Areas Program, he completed a report on *Significant Bedrock Fossil Localities in Maine*. In it he identified and described twenty-six places in the state that he recommended "be preserved for continuing and future study by earth scientists.[9] Forbes' report and its dissemination provided yet another opportunity for the public to learn about the Trout Valley Formation and its plant fossils.

At the time of the report, Forbes noted that he and others had been studying the fossil plants in the Trout Valley Formation for about thirteen years. The importance of the area is reflected in the number of pages Forbes wrote about it (about five pages) compared to the pages devoted to each of the other twenty-five localities (most are one to two pages). Around fifteen species of plants had been discovered in the formation by 1977. Based on the studies that had been done up to the time of the report, Forbes wrote that "we imply an assemblage of plants that occupied a few square miles in a topographical status that is comparable to a modern brackish or freshwater marsh, surrounded by modest elevations of perhaps a few meters on which are perched scattered patches of different species."[10] And among them lived one plant not known from any other place in the world at the time—*Pertica quadrifaria*.

In early 1985 in the 112th Legislature of the State of Maine, a bill surfaced that was assigned House Paper (H.P.) number 222 and Legislative Document (L.D.) number 256: "An Act to Designate *Pertica quadrifaria* the Official Fossil of the State of Maine." It was one of more than 2,000 bills considered for the first session. The bill was cosponsored by Representative John E. Masterman from Milo and Senator Charles P. Pray from Millinocket to be submitted by the Department of Conservation.

The Legislative Record shows that on January 29, 1985, Representative John Lisnik from Presque Isle presented the bill to the Maine House of Representatives. The bill was sent up for "concurrence"—meaning agreement between the House and Senate.[11] The next day the bill came to the Maine Senate and was referred to the Committee on State Government and ordered printed in concurrence.[12] What is not found in the Legislative Record is that the bill stimulated an extended period of deliberation about what would best qualify as the state's fossil.

The Committee on State Government, according to rule, was made up of three members of the Senate and ten members of the House. If anything can be said about the committee's membership, it was diverse—in values, life experiences, ages, careers, and political philosophies. Four were female, most members had attended college, and at least three were veterans. Its members were educators, consultants, researchers, and planners. Their interests ranged from the arts—poetry, music, theater—to industrial and real estate development. At least two were educators and one sold vehicles. As one would expect, they were involved in local and statewide affairs, community development, public administration, and a wide range of private organizations.[13] With this kind of mix, no one could predict the outcome.

On February 14 a public hearing was held by the Committee on State Government to learn how the public felt about having a state fossil. But before the state fossil bill came up, the panel heard speakers on a bill for an official state cat—the coon cat. The testimony caused some thought among the panel members about the trend of making official state symbols out of things that did not originate in Maine and led to a discussion about whether or not the coon cat was a native of the state. Remarks at the coon cat hearing were, at times, colorful and humorous, keeping the mood upbeat and generating good feelings. Dan Lambert, a well-known and highly respected decorated war hero, then adjunct of the American Legion of Maine, told the committee: "If you want to keep the state from going to the dogs, then adopt the coon as the state cat."[14]

Following the testimony on the coon cat, the committee took up the bill for the state fossil, hearing arguments for *Pertica quadrifaria*. William Forbes "testified that the fossil is one of the most unique in the world . . . [and] allows us to peek back through time at life as it was starting to emerge on the continents of the planet. Professor and geologist Stephen Pollock of the University of Southern Maine said the only place in the world where *Pertica quadrifaria* is found is in Baxter State Park."[15]

Another geologist who gave testimony was Walter A. Anderson, Director of the Maine Geological Survey. He pointed to "a growing trend among states across the nation to designate official state fossils. One of the major reasons for this trend is the desire among policy makers and scientists alike to draw attention to and to better acquaint the public with their state's geologic setting. This is particularly important in light of the significance of a variety of geologically-oriented issues today, such as hazardous waste management, mineral exploration, earthquakes, and landslides. I believe that Maine should join this trend." Anderson concluded with a few facts that made the fossil "an ideal candidate"—its importance in the evolution of vascular plants, including the state tree, the white pine; its exclusiveness; and its international significance. "Therefore," he said, "I urge you to support its designation as Maine's State Fossil."[16]

On February 20, a few days after the hearing, the committee received a letter opposing the designation of *Pertica quadrifaria* as the official state fossil. The writer's major point of criticism focused on a geologist's testimony that the fossil is "one of the most unique in the world." The writer then went on to say: "In considering any state symbol, uniqueness or rarity should not be an attribute of said symbol." He pointed out that other of Maine's state symbols are found in many places, such as the chickadee, landlocked salmon, pine tree, and the mineral tourmaline. "Some other fossil should be considered for a state fossil. Namely one not so rare.... A better choice would be one that is listed in the 'Audubon Society Field Guide to North American Fossils' namely Monograptus. This would certainly have a more popular appeal and be familiar to more people in all the States of the Union."[17]

Monograptus is a genus of graptolites, extinct aquatic colonial animals. They are small in size, and some species have been compared to coping saw blades in appearance. Graptolites appeared in the Silurian Period—444 to 419 million years ago. It is unclear whether or not the committee would consider a graptolite as being more popular and familiar, but usually the fossil of an animal has more of a chance for a selection than a plant fossil. Today, forty-three states have official state fossils, but only a half dozen or so are plant related. On this basis, the coon cat would seem to have a better chance in the attempt to designate a state cat.

On February 28th, despite uncertainty of the bill's chances, the Committee on State Government reported that the bill "Ought to Pass," and it was sent on its way through the traditional gauntlet of legal procedures.[18] Two weeks later on March 14, 1985, the *Kennebec Journal* announced that "Maine has an official state tree, animal, bird, flower, insect, fish and mineral. And now Gov. Joseph E. Brennan has signed a bill to give the state an official fossil.

"The *Pertica quadrifaria* is the prehistoric ancestor of the white pine cone and tassel, Maine's official flower, and the white pine tree, the official tree."[19]

On September 19, 1985, the law that made *Pertica quadrifaria* the official fossil of the State of Maine became effective. However, the fossil's story was not over, and scientists continued to study and write about locating new sites uncovered by the erosional effects of Trout Brook. Then, unexpectedly, a new opportunity occurred in the first years of the next century that led to additional studies of *Pertica quadrifaria* in the Trout Valley Formation.

The opportunity was in the Scientific Forestry Management Area (SFMA) of Baxter State Park, which had become part of the park in 1955, the year Rankin began his dissertation study of the Katahdin area's

Figure 9-1. Governor Joseph E. Brennan in March 1985, signing the legislative bill designating *Pertica quadrifaria* as the Maine State Fossil. Behind him looking on, from left to right, are Representative John Lisnik, Walter A. Anderson, Director, Maine Geological Survey, and William H. Forbes, Maine State Paleontologist. Photographer unnamed.

geology. Part of the area was underlain by the Trout Valley Formation where the erosional effects of Trout Brook had uncovered fossils. But beyond the brook the area was covered by dense forest vegetation and soil. Henry Andrews just knew that, if he could remove the forest cover, he would find more fossils. However, in Baxter State Park there were restrictions to digging and upsetting the wilderness character of the area. And so it had not been thought possible to dig and uncover new fossil localities, that is, until 2004.

10

NEW REVELATIONS

In 1988, three years after *Pertica quadrifaria* officially became the Maine State Fossil, the four principals in the discovery of the major Devonian fossils in the Trout Valley Formation—Andrew E. Kasper, Jr., Patricia G. Gensel, William H. Forbes, and Henry N. Andrews, Jr., published a review paper summarizing plant paleontology in Maine to that date.[1]

They noted, that within the past two decades (from the 1960s to the 1980s), our knowledge of plant life in the Devonian of Maine and North America vastly increased and, in Maine, the most significant information from fossil-plant studies had come from an area "long believed barren of any useful [fossil] information"—Trout Brook and its tributaries in the remote wilderness of Baxter State Park.[2]

Twelve plant species in all had been discovered in the Trout Brook Formation, more forms than from any of Maine's other plant fossil localities. These represented some of the earliest vascular land plants. All lacked true leaves. Instead, photosynthesis was carried out in small spiny projections from the stems (enations) and from the main axis and branches. An important advance in Devonian plants was the development of a vascular system. This is the key evolutionary innovation that allowed plants to live on the land with a transport system for moving water and nutrients. And among the vascular plants discovered in the Trout Valley Formation, *Pertica quadrifaria* was considered the most impressive.[3]

In concluding their 1988 review of two decades of exploration and study of the Trout Valley Formation, the four scientists, in an act of prescience, wrote that evident to them "is the fact that repeated collecting at known outcrops and continued exploration for new plant-bearing strata are likely to yield even more data."[4] This was a prophetic insight, which came true in 1999 with the arrival of a new scientist on the threshold of a new century and new revelations in the Trout Valley Formation.

This was Robert A. Gastaldo, a geoscientist schooled in geology, sedimentology, botany, paleontology and newly appointed to the position of Whipple-Coddington Professor of Geology and Chair at Colby College in Waterville, Maine. Although in a new position in Maine, he was not new to geology and paleontology having held an Alumni professorship in geology at Auburn University in Alabama for more than twenty years. There he had built an enviable record of research and scientific reports (more than eighty—practically all in paleontology) along with many honors, awards, and grants.[5]

Gastaldo's college education had begun at Gettysburg College in Pennsylvania. He had set out to become a physician, enrolling in biology. But like so many others we have met in this story, he was captivated by fossils and drawn into studying them. While at Gettysburg he took an independent study course and came under the influence of professor William C. Darrah, a botanist, geologist, and paleontologist studying plant fossils in the coal fields of Pennsylvania. This and other experiences at Gettysburg developed in Gastaldo an interest in botany and geology that led him to graduate studies at Southern Illinois University at Carbondale and master's and Phd degrees in those fields. Today, he is considered an expert in the evolution of terrestrial plants and peat-forming environments, paleoecology, fossilization, extinction and recovery, sedimentology, stratigraphy, and deep time climate reconstruction.[6]

As a paleobotanist who studies the geologic past through fossil organisms, Gastaldo knows that only a small part of any population of organisms is recorded in the rocks, and these are most likely to be the most common or dominant at the time. Thus, paleontology only gives us, in his words, "small windows on biodiversity" during different periods of geologic time. They are like "trailers for movies," he explains. And, he notes that what we see in the rock record accounts for only the last

Figure 10-1a. Robert A. Gastaldo. Photo taken in Xinjiang Province, China. Courtesy Robert A. Gastaldo. Photographer unnamed.

moments of what happened before the fossilization process began. The farther we go back in time, the fewer analogs (representations) we relate to the present. But, from these analogs we can learn about disturbances to populations such as extinctions and ecosystem responses. We can develop models that give insights to the past and understand problems over time.[7]

When he arrived at Colby College, Gastaldo also brought with him his interest in global environments of ancient coastal deltas. This led him to the Trout Valley Formation where, he believes, one of the earliest coastal ecosystems is preserved. He sees it as a classic and first principal location: rare and unusual in its diversity and the quality of its preserved fossils. In his words, "It's a golden spike."[8]

At Colby, Professor Robert E. Nelson in the Geology Department joined Gastaldo in his interest in the Trout Valley Formation. Nelson had a Ph.D. in multidisciplinary Quaternary paleoecology and had been in the Geology Department since 1982. That year he had become a John D. and Catherine T. MacArthur Assistant Professor and, by 1996, he had become chair and full professor. When Gastaldo joined the department,

Nelson had a publishing record of numerous abstracts and more than thirty scientific papers, either authored singly or with others.[9]

By the end of 2002, the two scientists had applied for and were awarded more than $100,000 from the National Science Foundation for a study revisiting the work of Andrews, Kasper, and others on the Trout Valley Formation.[10] In the meantime, Gastaldo and Nelson had begun supervising undergraduate students in field-laboratory studies related to the Trout Valley Formation. These included investigations into sedimentary rocks and plant-animal fossils, and they had published abstracts of their work.

One of the papers noted that earlier studies of the Trout Valley Formation lacked a detailed discussion of its strata and sedimentary context from which a model for deposition could be developed. Such a study was carried out in the summer of 2001 and proposed a sedimentological framework of the Trout Valley Formation.[11] In another paper, researchers in that same year 2001 reported that preliminary field investigations of outcrops along Trout Brook and South Branch Ponds Brook revealed that fossil plants had been transported short distances from where they grew. The transported plants were buried in tidal channels or in beds of sediment sloping down the front of an ancient delta (called an accretionary foreset). But, the paper concluded that more information was needed to confirm earlier interpretations of what was in the Devonian plant community.[12]

A third study was conducted by Nelson, Gastaldo, and others in the summer of 2001 "in hopes of recovering arthropod remains." Arthropods are invertebrate animals having an external skeleton (exoskeleton), a segmented body, and jointed appendages. They include insects, spiders, and crustaceans amongst other forms. Most previous studies of the Trout Valley Formation had identified vascular plants, but very little evidence of arthropods was found. Researchers in the 2001 study recovered arthropod remains, but they were very rare—"one arthropod fragment per several thousand plant fragments." They also collected samples of rocks bearing *Pertica* fossils.[13]

A report of eurypterid remains was published the next year in 2002 by Terkla, Nelson, and Gastaldo. Eurypterids are an extinct group of arthropods, known as sea scorpions. Rock samples were collected from twelve sites along Trout Brook, including two representing an ancient estuary and braided-stream plain. Fragments of two eurypterid families were found, encouraging the researchers to do further work.[14]

Beavers played a role in a discovery of a heretofore unknown siltstone outcrop of the Trout Valley Formation in the Trout Brook area. The outcrop was exposed by low water caused by a beaver dam upstream. In 2005 a study of the outcrop was reported, supported in part by the NSF grant to Gastaldo and Nelson. The researchers had discovered an estuarine assemblage of fossils consisting of a juvenile eurypterid, small gastropods, an ostracode (a small crustacean called a "seed shrimp" with a bivalve-like shell), and plant fragments. Evidence indicated that the animals lived in a near-shore estuarine tidal flat or channel.[15]

It is rare to find such assemblages and they are important because they help us understand an ecosystem in the Devonian. However, such an understanding is dependent on scientists who have honed their skills to make keen observations, who can search out what others have reported, and who can develop meaningful interpretations based on the evidence they find. For example, Gastaldo and Nelson (2005) knew that the species they found were typical of tidal flats and channels. They also suggested that the fossils in the assemblage had not been transported by water currents to where they were found because some were not fragmented. The presence of the nearly complete juvenile eurypterid also suggested it had not been moved any great distance from where it had lived. And, the plant fragments tied the assemblage to a nearshore environment, such as an estuary, and their orientation in the rock pointed to their being deposited in calm water following a high energy storm directed landward.[16]

In 2006 Jonathan Allen and Gastaldo were co-researchers on another project that studied the characteristics of the plant-bearing sedimentary strata in which *Pertica quadrifaria* and other plants of the Trout Valley Formation are preserved. This information was necessary to test two hypotheses that Kasper and Andrews had made: (1) *Pertica quadrifaria* and its associated plants grew in a land-based brackish (slightly salty) marsh, and (2) their fossil remains were found where they lived (not transported by moving water from some distant place).[17]

Allen and Gastaldo (2006) identified eighteen vertical sections of rock along Trout Brook, Dry Brook, and South Branch Ponds Brook. They photographed all exposed outcrops and collected hand samples of all sedimentary and plant-bearing rock for laboratory analysis. In the field they identified the kind of sediments (lithology) and bedding structures and features of the sediments, such as cross-bedding and ripples, and noted any disturbance of sediments. Samples with fossils were split along bedding planes to analyze how the organisms became fossilized. Thin-section analysis of the rock types (a 30-micrometer-thick sample ground transparent) was used to study both sedimentary rocks and fossils.[18]

The researchers identified seven different environments of deposition of rock (called facies) based on their sedimentary and fossil characteristics. In their discussion, the authors noted that Andrews, Kasper, Gensel, and other paleontologists in the 1960s and 1970s, in their extensive investigations of plant morphology and anatomy, interpreted the Trout Valley plants as living in a terrestrial brackish or freshwater marsh surrounded by land of modest elevation. The plants were of low diversity and preserved where they grew. However, these new studies found that the Trout Valley Formation represented a range of continental and nearshore marine environments. They ranged from a braided river plain, a flood-plain and estuary, a terrestrial environment with a poorly developed ancient soil horizon containing evidence of plant roots, and more open marine environments. Well-preserved plant remains showed that plants grew primarily in two environments—fluvial (river or stream) and estuarine/tidal. In these environments, Allen and Gastaldo found only one assemblage that had been preserved where it grew. All other fossil assemblages had been transported. But the authors believed that many fossil-plant communities must have been in their original setting to have had so many well-preserved plant stems.[19]

The larger trimerophytes, such as *Pertica quadrifaria*, were found in coarser sediments suggesting that they grew near the margins of active stream channels. These plants were transported during storms and flood waters before fossilization. The occurrence of smaller trimerophytes in finer sediments above a rooting layer of an early soil suggests the plants may have also lived in a coastal flood basin and wetland marsh. However,

Figure 10-1b. Robert A. Gastaldo with Patricia G. Gensel fossil hunting in Trout Brook area. Gensel was among the earliest group to study the Trout Valley fossils, beginning in the 1960s, and Gastaldo was in the latest group of fossil hunters, publishing a paper in 2016. Photo courtesy of Patricia G. Gensel. Photographer unnamed.

the researchers found no evidence of whether or not plants could tolerate brackish conditions. They did find evidence that trimerophytes had roots and pointed out that the roots originated in an organically-rich soil horizon.[20]

Over and above, what this study brought to our understanding of the Trout Valley Formation and its early plants, including *Pertica quadrifaria*, is a new level of how science moves forward. Science builds on previous studies—adding new information, correcting assumptions, confirming or disproving hypotheses, and leading to new ideas to be tested. The work of science is never done. Nor should it be. In the story of *Pertica quadrifaria*, after the work of Andrews, Kasper, and the other scientists of the 1960s and 1970s, one might think that the research in Trout Valley was pretty much done. But it wasn't!

If any of those first fossil hunters had a dream about that Trout Brook area, it was, perhaps, best expressed in their 1988 review paper. Here, they write: "It is tantalizing to speculate on what fossils lie hidden in the surrounding heavily wooded areas."[21] And speculate they did, but that was all they could do—speculate. The rules of Baxter State Park are stringent about removing or altering its natural character. It was legislated to be a wilderness and Governor Baxter made sure that it always would be so by his deeds of trust. However, something happened that made the speculation of the fossil hunters not so hopeless after all. But, sadly for Rankin, Dorf, Andrews, and Forbes, it was too late: they were gone. That "something" had begun to happen years earlier in 1955 when Governor Baxter had added the Scientific Forestry Management Area (SFMA) to the park. This was the year that Rankin was just beginning his study which first revealed the fossils that would eventually lead to the discovery of *Pertica* years later. Through those years, the park had administered the SFMA around Trout Brook as a demonstration of how to manage a forest for sustainable tree harvesting and providing to the citizens of Maine other benefits. Part of this effort was the careful planning and construction of roads.

It was in the year 2004 that the Baxter Park Authority approved road extensions in the SFMA. One of those extensions ran for a short distance parallel to, and about a quarter of a mile north of, the park's Tote Road. At this point, a short road was built to connect to the Tote Road.[22]

It was in this vicinity that the fossil hunters' dream of what fossil treasures might lay beneath the floor of northern Maine's dense boreal forest began to come true. It was here that a bulldozer, building the road extension, exposed fossiliferous rock. And it was here that Gastaldo and his students could begin a new phase of fossil studies in the Trout Valley Formation. Collecting took place in 2008, 2011, and 2014. As a result, thirteen sites with fossils along a transect near the road were identified from hand samples taken for studies similar to their previous work.[23]

The researchers examined several different facies: sandstone, sandy siltstone, and siltstone. The siltstone facies contained the largest concentration of plant, macroinvertebrate, and trace fossils. Plant fossils occurred in nearly every collection site and consisted primarily of roots

Figure 10-2. Fossiliferous rocks on the edge of a bulldozed road extension in the Scientific Forestry Management Area (SFMA) in Baxter State Park. Photo by Dean B. Bennett.

and stems. Plants were preserved in a variety of modes including compressions, impressions, and sediment-cast stems. And again *Pertica* was one of four genera identified. It was preserved in sandy siltstone and siltstone facies and fossils were buried both near where the plants had grown or to where they had been transported. At the east end of the transect at the high end of the ancient strata, Gastaldo (2016) reported plant fossils with vertically oriented roots preserved where they once lived in a paleosol, an ancient soil horizon.[24]

After reviewing all the data from the study, Gastaldo concluded that the bedrock along the Wadleigh Mountain Road preserved fossils in sediments deposited in a coastal estuarine paleoenvironment. This area likely was a shallow, brackish water, estuary. It is the most common environment in which storm-produced floods transported plants and sediments, burying them in successive layers that eventually turned the sediments into rock and the plants into fossils. Also buried and subsequently preserved in the sedimentary rock were animals, including the

Figure 10-3. A fossil plant in a rock from a bulldozed roadside in the SFMA in Baxter State Park. Photo by Dean B. Bennett.

mollusk *Modiomorpha* and the associated trace fossil *Spirophyton* (as the name implies, a spirally organized trace-fossil). When all was said and done, the study extended the area where fossils have been recovered from the Trout Valley Formation, including *Pertica quadrifaria*.[25]

When asked how his work had added to that of his predecessors who studied the fossils of the Trout Valley Formation, Gastaldo answered that it added "context." He had shown that *Pertica* existed in a wide range of habitats and physical conditions. For example, the rocks exposed with their fossils in the South Branch Ponds Brook area are made up of tidal deposits. However, *Pertica* is also thought to have existed in fresh-water marsh settings. Interestingly, Gastaldo noted that charcoal has been found in strata in the Trout Valley Formation, showing evidence that fire also operated as a physical factor in these ecosystems.[26]

In revealing the secrets of the Trout Valley Formation, hidden in the rocks of northern Maine's deep forest, researchers have developed a story that embraces the science of paleontology, the evolution of life on earth, and a monumental event in the history of our planet—the greening of

Figure 10-4. The ancient plant, *Pertica quadrifaria*, is depicted in its habitat 398 million years (mya) ago when it lived south of the equator. The plant in this illustration is based on a study of the plant's fossil by Andrew E. Kasper, Jr., and the scientific artistry of Mary Hubbard, staff artist of the Biological Sciences Group, University of Connecticut, Storrs. The depiction of the plant's habitat was designed by Dean B. Bennett with Google Earth and Photoshop and drew on interpretations made by Andrew E. Kasper, Jr., and Robert A. Gastaldo, who both studied the fossil plant extensively. Sources: Andrew E. Kasper, Jr., and Henry N. Andrews, Jr., "Pertica, A New Genus of Devonian Plants from Northern Maine," *American Journal of Botany*, vol. 59, no. 9, October 1972, pp. 897–911, and Robert A. Gastaldo, "New Paleontological Insights into the Emsian-Eifelian Trout Valley Formation, Baxter State Park's Scientific Management Area, Aroostook County, Maine," *Palaios* v. 3 (2016): 339–340.

the world. And we now know that a star in this story, once standing tall on the landscape above its other relatives, was *Pertica quadrifaria*. Like the celebrities of theater and film who left their footprints and handprints in the concrete in front of Grauman's Chinese Theater on Hollywood Boulevard, *Pertica* left its own mark embedded in the rock. But in the case of *Pertica*, its prints are all that are left. Its species and associated fellow trimerophytes are all extinct, whereas the species of the Hollywood stars are extant.

Pertica quadrifaria and its trimerophyte relatives, however, can be thought of surviving in their descendents: the progymnosperm (an extinct pre-gymnosperm group of the later Devonian) and living gymnosperm (an extant group of cone-bearing trees of which the white pine is one). Much simplified and in broad terms, the trimerophytes gave rise to early pre-gymnosperms: a group of leafless shrub-size plants with an innovation in conducting tissues: wood. You will recall that the trimerophytes, relatives of *Pertica quadrifaria*, had conducting tissues: xylem for water and phloem for sugar transport. The xylem in the trimerophytes was minimal (called primary xylem). Wood is called secondary xylem, more abundant in amount, still transporting water but with an additional new function: mechanical support. Wood permitted the shrub-size plants to be taller than their trimerophyte predecessors. These early pre-gymnosperms gave rise to later pro-gymnosperms at the end of the Devonian: giant tree-size plants with massive wood. With wood several feet in thickness, tall trees produced forests in Late Devonian time. The last step in this rough evolutionary sequence was the development of seeds. The tall trees of the Devonian forests, species of the genus *Archaeopteris*, were spore-shedding plants devoid of seeds. These pre-conifers of the Mesozoic gave rise to true conifers of which *Pinus strobus, the white pine*, is one. And in Baxter State Park it is growing on the remains of its far–distant ancestor, *Pertica quadrifaria*. A more detailed and thorough account of this evolutionary sequence is presented in the next chapter.

11

THE DESCENDENT

There is a hypothesis that *Pertica quadrifaria*, the Maine State Fossil, is an ancestor to the Maine State Tree, the eastern white pine, a conifer whose cone and tassel are the state flower. But is there scientific evidence that supports this? One needs to follow *Pertica*'s evolutionary pathway to attempt to answer that question.

Pertica quadrifaria is a species of an extinct land plant in the plant Division Trimerophytophyta. The trimerophytes emerged during the Early Devonian and came to an end in the Late Devonian, a span of time from 419 to 359 million years ago (mya). Like other early plants on Earth at that time, they were preceded by simple green algae, to which they were most closely associated as molecular evidence from the earliest rocks confirms. Algae are simple plants capable of photosynthesis, that is, with sunlight, water, and carbon dioxide they can produce food for themselves and most life forms on the planet. And, importantly, in the process oxygen is produced as a by-product.

Using evidence from 1) the plant's cell structure, and 2) its biochemical pathways—a chain of chemical reactions, and 3) its DNA and RNA—molecules for cell functioning, scientists conclude that an ancient group of green algae named the Charophyceae or charophytes are ancestoral to the first land plants. They successfully colonized the land more than 500 mya, an event that led to our planet's amazing diversity of plants as well as changes in its atmosphere and other living conditions that support life on Earth.[1]

Following the algae, the most primitive group of vascular land plants, the Division Rhyniophyta, evolved. Rhyniophytes were small plants, about one foot tall with mostly equal forking stems. They had no surface outgrowths and bore a single sporangium at each stem tip. The sporangium is a capsule in which reproductive spores are produced and stored. In the center of each stem a small cylinder of xylem and phloem cells provided for transport of water and nutrients.

Enter now, the trimerophytes, of which *Pertica quadrifaria* was one. The trimerophytes were the link in the chain between the more primitive rhyniophytes and the more advanced pre-gymnosperm plants. The trimerophytes developed a number of crucial modifications in plant structures. 1) "overtopping" was a growth pattern in which a stem forked, producing two unequal limbs. This simple modification permitted the evolution of a main stem (dominant limb) with side branches (lesser limbs). This allowed for much greater height in trimerophytes, perhaps up to six feet, rather than the one-foot rhyniophytes. 2) Along with this new growth pattern, there developed more robust vascular tissue, xylem and phloem, which supported such height. 3) The "condensation or collapsing" of the much-branched side stems each with its sporangium resulted in large clusters or masses of sporangia in the trimerophyte branch ends. This was in contrast to the single sporangium at the stem tips in the rhyniophytes. 4) A critical innovation in the trimerophytes was the development of surface outgrowths, "enations" on the stems. An increase in stem surface area meant an increase in photosynthesis, which increased food production. These four modifications in the trimerophytes demonstrate the group's critical link in the transition and evolution from the rhyniophytes to the pre-gymnosperms and eventually to the gymnosperms and our native conifer, the white pine.[2]

The progymnosperms were an important component of the Earth's plant life from the Middle Devonian through the Lower Mississippian, 390 to 350 mya—overlapping the trimerophytes. The progymnosperms developed a major evolutionary advancement, the cambium. Cambium is a tissue that undergoes constant cell division producing other tissues. The progymnosperms produced two cambia: an inner one producing secondary xylem or wood and an outer one producing cork as part of the bark.

Both wood and bark are features of today's conifers. However, progymnosperms did not have seeds like present-day conifers, but instead, like the trimerophytes, they had spores for reproduction. Wilson Nicholas Stewart et al. write in their book *Paleobotany and the Evolution of Plants* that "when we compare the characteristics of progymnosperms with those of trimerophytes, it seems highly probable that progymnosperms also evolved from this ancient and primitive group of plants."[3] This conclusion was also strongly supported a decade and a half later in 2009 by Taylor et al. in their critically important book *Paleobotany: The Biology and Evolution of Fossil Plants*: The progymnosperms "though poorly understood, provide the most convincing evidence of lineage ancestoral to seed plants."[4]

The connection between the trimerophytes and progymnosperms was strengthened when in 1975 a new fossil plant, *Oocampsa cathe*ta, was reported from New Brunswick, Canada. Two authors were Henry Andrews and Andrew Kasper, who discovered *Pertica quadrifaria*. A third author was Patricia Gensel, Andrews' former student who had described *Kaulangiophyton akantha*. Studies of *Oocampsa*'s spores showed that, though they were highly distinctive, they had features in common with the spores of gymnosperms.[5]

Within the progymnosperms are two orders, the "primitive," earlier Devonian, shrub-size plants with limited wood called aneurophytes (order Aneurophytales) mentioned previously and the "advanced" later Devonian, tree-size plants with massive wood called the archaeopterids (order Achaeopteridales). The latter order is based on the genus *Archaeopteris*—not *Archaeopterix* (the "dinosaur-bird")—and its name means "ancient" "fern." It was a tree over 80 feet tall and made up a late Devonian forest. The tree bore large fern-like branches and its wood anatomy was remarkably similar to that of present-day conifers, e.g., white pine.[6]

The botanist C. M. Govil wrote in 2007 that "there is general agreement that Archaeopteridales have a group of plants from which Coniferophytes [conifers] evolved." He also wrote that *Archaeopteris* exhibits characteristics which are found in the order Cordaitales and other members of Coniferophyta."[7] The Cordaitales are an extinct order of woody plants, true gymnosperms with seeds, thought to have descended from the progymnosperms 360 mya in the Carboniferous Period. They had reproductive cone-like structures bearing seeds suggesting cones of today's conifers.[8]

Conifers are a division of gymnosperms, mostly trees, with abundant wood, simple pollen cones (strobili), and compound seed cones. Their leaves are simple needles or scales, of which most are perennial—lasting many years. Conifers are very successful and today the coniferous forests, also known as the boreal forest and taiga biome, cover millions of square miles of the planet. There are more than 550 species of conifers in the world, one of which is the eastern white pine, *Pinus strobus*.[9]

The eastern white pine prefers a cool, dry climate and is primarily found from Newfoundland west across the Great Lakes region to south-eastern Manitoba and Minnesota and southward along the Atlantic coast to New York and inland along an Appalachian extension to Tennessee and Georgia.

In Maine, which is called the pine tree state, the importance of the white pine was recognized in 1820, the year that Maine became a state. It was then, at the first meeting of the Maine Legislature, that the white

Figure 11-1. The Maine State Flag featuring a white pine tree.

Figure 11-2. A white pine tree overhangs the site where its ancestor, *Pertica quadrifaria,* was first discovered along Trout Brook in Baxter State Park. Photo by Dean B. Bennett.

pine tree was adopted as one of the emblems in the shield of the State coat of arms which is now displayed prominently on the State flag. In 1895 the Maine Legislature declared the State Flower as the pine cone and tassel, and in 1945 the Legislature designated the white pine as the official State Tree.[10] Today, in Maine, the tree continues to be recognized throughout the state, making its presence known by its imposing size. And in Maine's and New England's largest wilderness area, Baxter State Park, a white pine overhangs the site of *Pertica*'s first discovery on the shore of Trout Brook (perhaps, poetically, in a sign of respect for its ancestor).

With some irony, *Pertica quadrifaria* might feel somewhat at home if it were alive today because studies have shown that it grew on the margins of river and stream channels. Here in the valley of Trout Brook near the fossilized remains of *Pertica quadrifaria*, the white pine is also at home because it, too, grows in moist areas—stream banks, river flats, and wetlands, and on sandy soils. But unlike *P. quadrifaria*, white pine also grows on hills and in mountains. However, it grows best on fertile, well-drained soils and in sandy areas where it can establish nearly pure forests.

Today the white pine stands tall on the landscape like *Pertica* once did, but unlike *Pertica*, the white pine came to be valued by humans. When Europeans appeared in the 1500s and 1600s, the white pine soon caught their attention, and through the years, its value was recognized in many ways, provoking both emotional and physical responses. Some saw masts for ships in the tall pines. It was in the mid-1600s that ships carried pine logs to the British Isles for the Royal Navy's use as masts. In 1691 the British Crown reserved for the King's navy all white pine within three miles of navigable waters and having a diameter of twenty-four inches a foot above the ground with a yard of height for each inch of diameter above the butt and which stood on land not previously granted to private individuals. A surveyor-general's position or office was also established to blaze chosen trees with a marking hatchet by making three cuts through the bark of the tree in the shape of an arrowhead—the sign of Royal Navy Property. This mark came to be known as the King's Broad Arrow, and as far as we know, no evidence of it has been found in the vicinity of *Pertica*'s fossils in the East Branch country.[11]

Others saw something entirely different in the white pines. Henry Wadsworth Longfellow, an American celebrity poet, wrote in 1847 his epic poem "Evangeline" with the memorable words: "This the forest primeval. The murmuring pines and the hemlocks." The pines were the eastern white pines and the hemlocks, the eastern hemlocks.[12]

A contemporary of Longfellow was the naturalist Henry David Thoreau, whose writings included many observations and thoughts about the white pine. For example, in his essay "Walking," he wrote about climbing "a tall white pine on the top of a hill, and though I got well pitched I was well paid for it, for I discovered new mountains in the horizon which I had never seen before,—so much more of the earth and the heavens. I might have walked about the foot of the tree for three score years and ten, and yet I certainly should never have seen them."[13]

From 1846 to 1857 Thoreau made three trips to the Maine woods. Unlike Rankin, who had come to decipher the rocky underpinning of the country, Thoreau was a naturalist who had come to see a wilderness. Lumbermen had been in the Penobscot River's East and West Branches ten years before Thoreau made his first trip. It was in this country in 1857 that Thoreau, heading toward the East Branch of the Penobscot, canoed by the mouth of Trout Brook, less than a mile from the logging depot Trout Brook Farm where Douglas Rankin would stay a century later.

His Maine woods experiences left him disturbed by the logging he saw, especially the cutting of the big pines, for during the time when Thoreau was making his three trips to the Maine woods, the white pine was Maine's most important timber tree.[14] It was a time of extensive cutting throughout the eastern parts of the country, and in just one year during this time, a quarter of a million white pines arrived in Chicago lumber yards.[15]

Before the Europeans arrived on this continent, it was estimated that the virgin stands of white pine held 600 billion board feet of lumber (a board foot is one foot long by one foot wide and one inch thick).[16] By the late 1800s, they were nearly gone, the result of the white pine's desirability for its many uses. In Maine's upper reaches of the Penobscot where *Pertica* quietly lay beneath the ground, the large virgin pines had been cut so heavily that they were hard to find, succumbing to the myth that they were so plentiful that they could not be depleted. But the waste and inefficiencies of harvesting, transporting, and reducing the trees to lumber had taken their toll.

By 1900, with changes in the forest industry and the need to have sustainable supplies of wood for larger businesses, entrepreneurs and the general public became increasingly receptive to the idea of conservation. It was this concept of sustainability that helped Governor Percival Baxter succeed in obtaining an area on the north side of Trout Brook, some 28,000 acres in size, and making it a part of his park. By then, he had already protected thousands of acres on the south side of the brook where *Pertica* had been discovered and protected it in a wilderness area. But Baxter had a different idea for this new area. In 1955, he wrote:

> *It has long been my purpose to create in our forests a large area wherein the State may practice the most modern methods of forest control, reforestation, and production. . . . I want this township to become a showplace for those interested in forestry, a place where a continuing timber crop can be cultivated, harvested, and sold, where reforestation and scientific cutting can be employed, an example and an inspiration to others. What is done in our forests today will help or harm the generations who follow us.*[17]

It was not until more than twenty years after he set forth his ideas for the Scientific Forestry Management Area (SFMA), as it became known, that his wishes began to be carried out. A park forester was hired in 1978 to begin the first draft of a scientific forest management plan. Three years later in 1981, the plan was approved by the Park Authority. The five-year plan's purpose was to develop the area into a "showplace," as Baxter had wished, an area to serve as an example and inspiration to those promoting scientific forestry. Harvesting was to begin and roads were to be built, designed for the least possible impact.[18]

In 1999 a visitor to the SFMA, John Herrick, reporting for the Forest Ecology Network, found that the area was, indeed, following a sound plan for "ecological forestry not forestry just for maximizing short-term profits." The plan, he said, followed basic precepts: "no clear cuts and no herbicides, cut the worst and keep the best, and cutting will not exceed growth." Herrick was also shown a section of old-growth forest in the SFMA that "will never be cut." There, he saw "an enormous healthy white pine which must have been 250 to 300 years old . . . less than one-tenth of one percent of Maine forests are old-growth." He concluded, "Thanks to the vision of Percival Baxter at least a little bit is preserved."[19]

Today, the SFMA continues to be managed for sustainability, and the white pine is among the species being managed. However, there lurks here a threat to the white pine and other species of plant life that is raising the concern of citizens throughout the world, including the managers of Baxter State Park and its SFMA. It is called climate change and if we are to meet its challenge and alter its devastating course, a course for which we must take some responsibility, we will need a perspective on nature that includes a clear view of our place in the natural world and responsibilities to it. That perspective must include an understanding of time, and in the case of climate change, what we can learn from the past that will help guide our decisions influencing the future. And it is the study of extinct plants and fossils like *Pertica quadrifaria* that will help answer these questions.

12

A PERSPECTIVE ON SCIENCE AND NATURE

ANY PERSPECTIVE ON SCIENCE AND NATURE MUST INCLUDE THE CONcept of conservation in the management of land, water, atmosphere, energy resources, and living things for the ecological stability of the planet. We have seen examples in this story of the Maine State Fossil and the place where it is found—Baxter State Park. The Park is a place where nature is protected in areas managed for wilderness and for sustainable use. Both areas have clear constraints on the use and alteration of Trout Valley and its fossilized plants.

The park has been identified as one of more than 140 areas in Maine designated as Focus Areas of Statewide Significance. These areas contain unusually valuable concentrations of at-risk species and habitats that have been identified by scientists of state, federal, and private conservation organizations. The goal is to encourage awareness, appreciation, and sustainable conservation of these exceptional places and other similarly important adjacent land areas. It is to be noted that, under the program, white pine forests in the park have been ranked as ***secure*** at both state and global levels. One of the program's conservation considerations identified with the park is climate change over the next century. The program also noted that, as the climate warms, alpine areas in the park "may become important as refuges for native species and communities currently found at lower elevations."[1]

A paper prepared under the auspices of the Park reported that "climate change has emerged as one of the most urgent and important issues of this century." Despite the considerable amount of scientific research on the issue and the uncertainty of predicting local changes, "climate change has the clear potential to significantly, perhaps drastically, change the arrangement and interconnections of plant and animal communities, . . . the presence of species and the recreational and economic opportunities available to Park visitors in particular and society in general." The paper concluded that the best approach would be to monitor and identify possible risks and problems caused by climate change and to prepare to adapt to them. The most important adaptation would be to change existing management systems to meet the problems of a more rapidly changing environment. This would include effectively monitoring the changes occurring in the Park's environment and adapting management policies to meet identified problems.[2]

Adapting is one major way we can respond to a warming climate. Another way is to promote reforestation and sustainable use of forests. The International Union for the Conservation of Nature (IUCN) reports that forests help stabilize the climate by absorbing carbon dioxide. Preventing the loss and degradation of our natural systems can potentially contribute over one-third of what is required by 2030 to mitigate climate change.[3]

Plants and climate mutually influence one another, as was demonstrated more than 400 million years ago when plants invaded the land. We need a "regreening" of the planet today to make it more habitable by stabilizing the climate. And to help in the effort we are using fossils of ancient plants to better understand what is happening today. Studies of fossil leaves by paleobotanists help them understand ancient forests, past climates, and climate change. For example, studies of fossil leaves in Wyoming's Big Horn Basin, dating back to 56 mya, help scientists and people throughout the world understand and prepare for a surge in global warming. These fossil leaves are being studied to determine the effects of a sudden period of warming.[4] Fossilized leaves from Greenland and Denmark are also helping us to understand how our activities influence the global climate today and how we will influence the diversity of living things on the planet tomorrow.[5] Fossil plant stomata on leaves of extinct trees in central Germany growing 40 mya show decreasing carbon dioxide in the atmosphere.[6] Stomata are minute pores in the outer layer of a leaf that allow gases in and out.

These studies permit a rough idea of ancient forests and changes in forest makeup; paleobotany and paleo-climatology help us understand the time framework and frequency and severity of climate changes; these ideas are then used by present-day ecologists making theories and applications to present-day forests and climate change. The studies also help ecologists draw broad conclusions regarding the ecological impact of climate change; ideas and concepts generated by ecologists are then used by professional forest managers to develop strategies for good forest management and used by politicians and the public to develop good legislative policy.

Fossils also have a different but important use: they can help motivate children and adults to become interested in science and, for some, to become scientists themselves. We saw that happened to William Forbes, who overcame great odds to become a respected and highly accomplished professor of paleontolgy in Maine. And so, too, were fossils a motivator for one of America's best-known women scientists and ecological heroine—Rachel Carson. She became a research biologist when it was rare for women to do so and displayed a talent as a writer who could describe the natural world with both a poetic sense and a clear scientific understanding. But perhaps even more importantly, she stood out as a defender of our ethical obligations to the natural world.

Linda Lear, Carson's biographer, who also considers herself "a child of the Allegheny" region where Carson grew up, said that she "always had a hunch about the origins of [Carson's] passion for understanding nature and the science of biology and ecology." Carson's home, Lear wrote, was situated above the Allegheny River in western Pennsylvania where she discovered a fossilized fish in nearby cliffs. The girl's curiosity was stimulated, and she wondered if a sea had once been there. Later when she was in college, Carson was asked by her college classmates why she had decided to become a biology major. In response, she told them the story of finding the fossil. Lear wrote that Carson's "desire from the outset was to understand the life that had once lived in the past and its relationship to the present. It was always about water."[7]

Figure 12-1. William H. Forbes as a boy developed a passion for rocks and fossils that endured through his entire life. Photo courtesy of Warrena Forbes.

Rachel Carson's passion for the natural world was instilled by her mother and aided by Anna Botsford Comstock's book *Handbook of Nature Study*, first published in 1911 when Carson was four years old. It was noted by the writer Linda Engelsiepen that "one of Comstock's lessons was that 'nature's laws are not to be evaded'—a truth that would inform Rachel Carson's life and legacy."[8]

Like most scientists, Rachel Carson was ethically aware of our obligations to the future of the planet and its life, and she powerfully demonstrated this to the public through her book *Silent Spring* and her advocacy for a cleaner, toxic-free environment. So, too, in their own ways did the scientists we have introduced in this book demonstrate this love for nature and a sense of stewardship. Henry Andrews, for example, in addition to his productive research and writing, found time to actively volunteer for conservation work with the Lakes Region Conservation

Figure 12-2. William H. Forbes was a teacher as well as a scientist and throughout his life helped others understand the importance of understanding the world around them. Photographer unnamed.

Trust in New Hampshire.[9] And Bill Forbes inspired students in his teaching and did work for Maine's Critical Areas Program to identify fossil localities in Maine to encourage their conservation.[10] On May 3, 2011, at eighty years of age, he left us but not before having passed on his love of science and learning to hundreds of students and setting an example of the value and role of science in society.

All of the scientists we have encountered in this book searched for the truth in nature, that is, what is in accordance with fact and reality. This is a fundamental goal of science, a human effort driven and advanced by curiosity. This is behind the work of the paleontologists we have seen in this book, and for many, it started with fossils, perhaps in a remote wilderness outcrop along a mountain stream or in a roadcut along a well-traveled highway.

EPILOGUE

THE PALEOBOTANISTS IN THIS STORY, AFTER YEARS OF SEARCHING AND study, found the fossils of about a dozen vascular plants that grew in what is now known as Baxter State Park during the Devonian Period of time, about 400 million years ago. All were discovered in the Trout Valley Formation, named for the brook that flows through and over it. Their explorations, beginning in the 1950s, are recorded in this book through to the year 2016. In that year, another study of vascular plants growing in Baxter State Park was completed and published.[1] This study focused on the living vascular plants, and the scientists and naturalists conducting it found 857 vascular plants growing in the park that had evolved over the past 400 million years or more.[1]

We now know much about these plants, both ancient and current. What we do not know is what the future will hold for these plants and others growing in the park as we consider the effects of global climate change, human population growth, technological development, and changing human values and behavior, for example. But what we do know is this: because we and other animal life in the world are fundamentally dependent on the health of our plant life, it behooves us to exercise a responsibility to care for the planet. In doing this, the role of science will be critical.

NOTES

Prologue

1. Roy R. Lemon, *Vanished Worlds: An Introduction to Historical Geology* (Dubuque, Iowa: Wm. C. Brown Publishers, Inc., 1993), 13.
2. Robert Gastaldo, to the authors, March 2020.
3. See Christopher R. Scotese, *Paleomap Project*, Website: www.scotese.com/earth.htm

1. A Fateful Meeting

1. The story of this meeting in a road cut appears in several publications, for example: Ruth Mraz, "Washburn Man Termed Expert in Specialized Science Field," n.p., n.d. (in possession of the authors); "Memorial for Bill Forbes to be held following geology meeting," press release, University of Maine at Presque Isle, 6 April 2012; and "William H. Forbes," (obituary) *Bangor Daily News*, 03 May 2011.
2. Mraz, "Washburn Man Termed Expert."
3. Henry Andrews, *The Fossil Hunters: In Search of Ancient Plants* (Ithaca, New York: Cornell University Press, 1980), 221.
4. Walter Anderson, interview with the authors, Gray, Maine, 6 March 2018.
5. "Memorial for Bill Forbes."
6. Percy Edward Raymond, "Obituary. Olof O. Nylander [1894-1943]." *American Journal of Science*, 241, no. 11 (November 1943), 704–705; and Nylander Museum website, www.nylandermuseum.org/History.htm
7. "Memorial for Bill Forbes," and Mraz, "Washburn Man Termed Expert."
8. Warrena Forbes, letter to the authors, 28 August 2017.
9. Ibid.
10. Mraz, "Washburn Man Termed Expert."
11. Warrena Forbes, interview by the authors, 17 August 2016.
12. "Memorial for Bill Forbes."
13. See David Chalmer Roy, "The Silurian of Northeastern Aroostook County, Maine," Thesis for the degree of Doctor of Philosophy (Massachusetts Institute of Technology,

1970), 23. See also Robert R. Shrock, "Memorial to Ely Mencher 1913-1978," AAPG (American Association of Petroleum Geologists), August 1979.
14. Warrena Forbes, letter to the authors, 28 August 2017.
15. Arthur J. Boucot, et al., *Reconnaissance Bedrock Geology of the Presque Isle Quadrangle.* Quadrangle Mapping Series No. 2 (Augusta, Maine: Maine Geological Survey and Maine Department of Economic Development, 1964), 12.
16. James M. Schopf, *Middle Devonian Plant Fossils from Northern Maine*, U. S. Geological Survey Professional Paper 501-D (Washington, D.C.: U. S. Geological Survey, 1964), D43.
17. Anderson, interview, 6 March 2018.
18. Mraz, "Washburn Man Termed Expert."
19. Andrew E. Kasper, Jr., et al., "Plant Paleontology in the State of Maine—A Review," in *Studies in Maine Geology* (Augusta, Maine; Maine Geological Survey, 1988), 113.
20. Ibid., 114.

2. Wading into the Devonian

1. Douglas W. Rankin and Dabney W. Caldwell, *A Guide to the Geology of Baxter State Park and Katahdin* (Augusta, Maine: Maine Geological Survey, Maine Department of Conservation, 2010), 67.
2. Douglas Rankin's exploration of South Branch Ponds Brook drew upon the following sources: Rankin and Caldwell, *A Guide*, 67–70; and Douglas W. Rankin, "Early Devonian Explosive Silicic Volcanism and Associated Early and Middle Devonian Clastic Sedimentation that Brackets the Acadian Orogeny, Traveler Mountain Area, Maine," in Lindley S. Hanson, ed., *New England Intercollegiate Geological Conference, 85th Annual Meeting: Guidebook to Field Trips in North Central Maine* (Dubuque, Iowa: Wm. C. Brown Publishers, 1995), 135–146.

3. Preparing to Pursue A Passion

1. See "Helen Barnes Whiting 1915," *One Hundred Year Biographical Directory of Mount Holyoke College 1837-1937*, Bulletin Series 30, no. 5 (Hadley, Mass.; Alumnae Association of Mount Holyoke College, 1937).
2. See *Blue Hen Yearbook, 1925*, student-produced yearbook (Newark, Del.: University of Delaware, 1925), 22.
3. Douglas W. Rankin, "Somebody Should Build a Bridge Over This River: A young trail crew's feat, 1953," *Appalachia* 63, no. 1 (winter/spring 2012): 86.
4. Robert D. Tucker and Peter Robinson, "Memorial to Douglas Whiting Rankin, 1931-2015," in *Memorials*, 44 (Boulder, Colorado: Geological Society of America, 2015), 27.
5. Rankin, "Somebody Should Build a Bridge," 83.
6. Mary Backus Rankin, "About the Author," in Douglas W. Rankin and Dabney W. Caldwell, *A Guide to the Geology of Baxter State Park and Katahdin* (Augusta, Maine: Maine Department of Conservation, 2010), 81.

7. John W. Hakola, *Legacy of a Lifetime: The Story of Baxter State Park*, for the Baxter State Park Authority (Woolwich, Maine: TBW Books, 1981), 269.
8. Mary Rankin, "About the Author," 81.
9. Philip W. Choquette, "Memorial to John Grant Woodruff 1898-1971," Geological Society of America. See: ftp:/rock.geosociety.org/pub/Memorials/vo3/Woodruff-JG.pdf.
10. John G. Woodruff, "Memorial to David Woolsey Trainer, Jr.," *Geological Society of America Bulletin* 77, no. 5 (May 10, 1966): 81–82.
11. Colgate University (Student Yearbook), *Salmagundi—1953* (Hamilton, New York: Colgate University, 1953), 49.
12. Colgate 1953 yearbook, Outing Club Report, 116–117.
13. Rankin, "Somebody Should Build a Bridge," 83.
14. Ibid., 84, 86.
15. Ibid., 86.
16. Tucker and Robinson, "Memorial," 27.

4. Facing a Peopled Wilderness

1. Douglas Whiting Rankin, "Bedrock Geology of the Katahdin-Traveler Area, Maine," Thesis for the degree of Doctor of Philosophy in Geology (Cambridge, Massachusetts, Harvard University, 1961),1–7.
2. John W. Hakola, *Legacy of a Lifetime: The Story of Baxter State Park*, for the Baxter State Park Authority (Woolwich, Maine: TBW Books, 1981), 180, 245, 255.
3. Mary Rankin, "About the Author," *in* Douglas W. Rankin and Dabney W. Caldwell, *A Guide to the Geology of Baxter State Park and Katahdin* (Augusta, Maine: Department of Conservation, 2010), 81.
4. See Roger LeB. Hooke and Paul R. Hansen, "Late- and Post-Glacial History of the East Branch of the Penobscot River, Maine, USA," *Atlantic Geology*, 53 (2017), 285–300.
5. John E. McLeod, *The Northern: The Way I Remember* [East Millinocket, Maine]: Great Northern Paper, [1981], 15.
6. See Gordon G. Whitney, *From Coastal Wilderness to Fruited Plain: A History of Environmental Change in Temperate North America from 1500 to the Present* (Cambridge, England: Cambridge University Press, 1994), 175–176.
7. Shirlee Connors-Carlson, *The Proudwood People: 1886-1986, 100 Years as a Way of Life* (Allagash, Maine: Town Crier, 1986).
8. A brief history of the Monument Line related to the East Branch of the Penobscot River area is in David Little, John W. Neff, and Howard Whitcomb, *Penobscot East Branch Lands: A Journey Through Time* (Portland, Maine: Elliotsville Plantation, 2016), 16–17.
9. Ibid., 20–21.
10. Rankin, Thesis, 8.
11. Philip T. Coolidge, *History of the Maine Woods* (Bangor, Maine: Furbush-Roberts Printing Company, 1963), 42, 50–51.
12. Hakola, *Legacy of a Lifetime*, 30.

13. See Richard W. Judd, *Aroostook: A Century of Logging in Northern Maine* (Orono: University of Maine Press, 1989), 62–63.
14. "The Evidence before the Committee on Interior Waters, on Petition of Wm H. Smith, Daniel M. Howard, Warren Brown, and Theodore H. Dillingham, for Leave to Build a Sluiceway from Lake Telos to Webster Pond," reported by Israel Washburn Jr., 134493 Telos Canal [1846], 1, Maine State Law and Legislative Reference Library, Augusta.
15. Hakola, *Legacy of a Lifetime*, 30; and Coolidge, *History of the Maine Woods*, 65.
16. Randall Probert, *A Forgotten Legacy* (n.p.: Randall Enterprises, Inc., 1998), 200–201.
17. Charles A. J. Farrar, *Farrar's Illustrated Guide Book to Moosehead Lake and Vicinity* (Boston, Mass.: Lee and Shepard, 1889), 195.
18. Rankin, Thesis, 10.
19. Little, Neff, and Whitcomb, *Penobscot East Branch Lands*, 33.
20. For information on the Maine Scientific Survey of 1861 see Ezekiel Holmes and Charles H. Hitchcock, "Preliminary Report upon the Natural History and Geology of the State of Maine: 1861," *Sixth Annual Report of the Secretary of the Maine Board of Agriculture: 1861* (Augusta, Maine: Stevens and Sayward, Printers to the State, 1861).
21. See "Penobscot East Branch In 1861. From the Diary and Letters of Alpheus Spring Packard." Edited by Kenneth A. Henderson. Pamphlet 3849, Special Collections Department, Raymond H. Fogler Library, University of Maine, Orono.
22. Rankin, Thesis, 11.
23. Ibid., 12.
24. Ibid., 6.

5. Mapping a Wildland

1. Douglas W. Rankin and Dabney W. Caldwell, *A Guide to the Geology of Baxter State Park and Katahdin* (Augusta, Maine: Maine Department of Conservation, Maine Geological Survey, 2010), 33.
2. Douglas Whiting Rankin, "Bedrock Geology of the Katahdin-Traveler Area of Maine," Thesis for the degree of Doctor of Philosophy in Geology (Cambridge, Massachusetts: Harvard University, 1961), 25.
3. Rankin, Thesis, 28–29.
4. Mary Backus Rankin, "About the Author," in Rankin and Caldwell, *A Guide*, 81.
5. Rankin, Thesis, 26.
6. Ibid., 202.
7. Ibid., 204.
8. Much of this history, with exceptions noted, appears in Rankin and Caldwell, *A Guide*, 30–33.
9. Lisa Churchill-Dickson, *Maine's Fossil Record: The Paleozoic* (Augusta, Maine: Maine Department of Conservation, Maine Geological Survey, 2007), 269.
10. Dwight C. Bradley, Robert D. Tucker, Daniel R. Lux, Anita G. Harris, and D. Colin McGregor. *Migration of the Acadian Orogen and Foreland Basin Across the Northern Appalachians of Maine and Adjacent Areas*, professional paper 1624 (Washington, D.C.: U.S. Department of the Interior, U.S. Geological Survey, 2000), 1–12, 35–38.

11. Rankin and Caldwell, *A Guide*, 32.
12. Robert D. Tucker and Peter Robinson, "Memorial to Douglas Whiting Rankin 1931-2015," in *Memorials*, v. 44 (Boulder, Colorado: Geological Society of America, 2015), 29.

6. A Curious Formation

1. Erling Dorf and Douglas Rankin, "Early Devonian Plants from the Traveler Mountain Area, Maine," *Journal of Paleontology* 36, no. 5 (September 1962), and see also Douglas Whiting Rankin, "Bedrock Geology of the Katahdin-Traveler Area, Maine," Thesis for the degree of Doctor of Philosophy in Geology (Cambridge, Massachusetts: Harvard University, 1961), 207–208.
2. See Sheldon Judson, "Memorial to Erling Dorf, 1905-1984," Geological Society of America.
3. Erling Dorf, "Climates of the Past," Princeton '55 NBC Educational Television Mudd Manuscript Library. https://blogs.princeton.edu/mudd/2015/02/climates-of-the-past/
4. Transactions of the National Association of Geology Teachers, Minutes of the Executive Committee Meeting, Held in New York City, November 17, 1963, "Citation of Erling Dorf for The Neil Miner Award, 1963."
5. Judson, "Memorial to Erling Dorf."
6. Dorf and Rankin, "Early Devonian Plants," 999.
7. Dorf, E., "Lower Devonian Flora from Beartooth Butte, Wyoming," *Geological Society of America Bulletin*, 45, no. 3 (June 1934): 425–440.
8. Dorf and Rankin, "Early Devonian Plants," 1002.
9. Ibid., 999.
10. Rankin, Thesis, 207.
11. See Andrew E. Kasper, Henry N. Andrews, and William Forbes, "New Fertile Species of Psilophyton from the Devonian of Maine," *American Journal of Botany* 61, no. 4 (April 1974): 339–341.
12. Lisa Churchill-Dickson, *Maine's Fossil Record: The Paleozoic* (Augusta, Maine: Maine Department of Conservation, Maine Geological Survey, 2007), 280.
13. Dorf and Rankin, "Early Devonian Plants," 999.
14. Ibid., 999–1000,
15. Ibid., 1000–1001.
16. Churchill-Dickson, *Fossil Record*, 255–296.
17. National Research Council (US), Committee on Responsibilities of Authorship in the Biological Sciences, *Sharing Publication-Related Data and Materials: Responsibilities of Authorship in the Life Sciences* (Washington, D.C.: National Academies Press (US), 2003).
18. Dorf and Rankin, "Early Devonian Plants," 999.

7. The Fossil Hunters

1. Tom L. Phillips, "Henry Nathaniel Andrews, Jr., 1910-2002," in *Biographical Memoirs* 88 (Washington, D.C.: National Academy of Sciences, 2006), 4.
2. Sergius H. Mamay, "Henry N. Andrews, Jr.: A Biographical Sketch," in *Review of Palaeobotany and Palynology* 20 (1975):4.
3. Tom L. Phillips and Patricia G. Gensel, "Henry Nathaniel Andrews, Jr. (1910-): Paleobotanist, Educator, and Explorer," in Paul C. Lyons, Elsie Darrah Morey, and Robert H. Wagner, *Historical Perspective of Early Twentieth Century Carboniferous Paleobotany in North America* (Boulder, Colorado: The Geological Society of America, Inc., 1995), 246.
4. Andrews, H. N., Unpublished biographical Notes provided by Membership office, National Academy of Sciences, 1975, p. 2, and quoted in Phillips, "Andrews, 1910-2002," 5–6.
5. Mamay, *A Biographical Sketch*, 4.
6. Elizabeth Cotsibas, "Tilton-Northfield BPW Names 'Lib' Woman of Achievement," *Laconia Evening Citizen*, 16 October 1979.
7. Wilson A. Taylor, "In Memoriam: Henry N. Andrews, Jr., Paleobotanist, Educator and Explorer, 1910-2002," *Bibliography of American Paleobotany, Paleobotanical Section* (St. Louis, Missouri: Botanical Society of America, 2002), n.p.
8. Phillips and Gensel, "Henry Nathaniel Andrews, Jr. (1910-)"; and Phillips, "Henry Nathaniel Andrews, Jr., 1910-2002," 10, 12, 13.
9. Henry N. Andrews, *The Fossil Hunters: In Search of Ancient Plants* (Ithaca, New York: Cornell University Press, 1980), 241.
10. Andrews, *The Fossil Hunters*, 241–243; and Robert M. Kosanke, "Memorial to James Morton Schopf, 1911-1976," *Geological Society Memorials*, 10 (Boulder, Colorado: Geological Society of America, n.d. (no date), n.p. (no page).
11. Phillips, "Henry Nathaniel Andrews, Jr., 1910-2002," 11.
12. Andrew E. Kasper, Jr., Patricia G. Gensel, William H. Forbes, and Henry N. Andrews, Jr., "Plant Paleontology in the State of Maine—a Review," *Maine Geological Survey, Studies in Maine Geology, Vol. I* (Augusta, Maine: Maine Geological Survey, 1988), 113.
13. James M. Schopf, *Middle Devonian Plant Fossils from Northern Maine*, U.S. Geological Survey Professional Paper 501-D (Washington, D.C.: U.S. Geologic Survey, 1964), D48.
14. Arthur J. Boucot, Michael T. Field, Raymond Fletcher, William H. Forbes, Richard S. Naylor, and Louis Pavlides, *Reconnaissance Bedrock Geology of the Presque Isle Quadrangle, Maine* (Augusta, Maine: Maine Geological Survey, 1964).
15. Dabney W. Caldwell, ed., *New England Intercollegiate Geological Conference (NEIGC) 1966 Guidebook: The Mount Katadin Region, Maine,* 58th Annual Meeting, Sept. 29–Oct. 1, 1966 (University of New Hampshire Digital Collection).
16. Henry N. Andrews, Andrew Kasper, and Ely Mencher, "Psilophyton forbesii, a New Devonian Plant from Northern Maine," *Bulletin of the Torrey Botanical Club* 95 no. 1 (January-February, 1968):1, 3, and 6.

17. Background information on Andrew E. Kasper, Jr., was from Kasper interview notes 7 March 1988, by one of the authors (DB) at Rutgers University, in possession of the authors; Notes from meeting with Warrena Forbes and Andrew Kasper, 17 August 2016, in Washburn, Maine, in possession of the authors; and Curriculum Vitae, Andrew E. Kasper, Jr., PhD, in possession of the authors; Andrew E. Kasper, Jr., to the authors, 1 April 2020.
18. See Andrew Kasper interview notes, 7 March 1988, in possession of the authors.
19. Andrew E. Kasper, Jr. to the authors, 1 April 2020.
20. See Patricia G. Gensel (http://bio.unc.edu/people/faculty/gensel/); "http://www.ipni.org/w/index.php?title=Patricia_G._Gensel&oldid=854041583"; and "About the Authors," in Patricia G. Gensel and Henry N. Andrews, *Plant Life in the Devonian* (New York, New York: Praeger Publishers, 1984), n. p.
21. Henry N. Andrews and Andrew E. Kasper, "Plant Fossils of the Trout Valley Formation," *Shorter Contributions to Maine Geology Bulletin* 23 (Augusta, Maine: Maine Geological Survey, Department of Economic Development, 1970), 3–16.

8. The Discovery

1. Andrew E. Kasper, Jr., interview with author (DB), Rutgers University, 7 March 1988.
2. Ibid.
3. Henry N. Andrews, Jr., journal entries, 17–18 July 1968, copies of journal excerpts in possession of the authors.
4. Kasper interview, 7 March 1988.
5. Henry N. Andrews, *The Fossil Hunters* (Ithaca, New York: Cornell University Press, 1980), 345–347.
6. Harlan P. Banks, "The Scientific Work of Suzanne Leclercq," *Review of Palaeobotany and Palynology* 14 (August, 1972):1.
7. See Kasper interview, 7 March 1988; Andrew E. Kasper, Jr., "A New Genus of Devonian Fossil Plants from Northern Maine," Dissertation for the degree of Doctor of Philosophy (Storrs, Connecticut, University of Connecticut, 1970), 14–17; Andrew E. Kasper, "Pertica, A New Genus of Devonian Plants from Northern Maine," *American Journal of Botany* 59 no. 9 (October, 1972): 900.
8. Ibid.
9. Ibid.
10. Kasper interview, 7 March 1988; Kasper Dissertation, 33–34; Kasper, "Pertica, A New Genus," 904–906.
11. Kasper interview, 7 March 1988; and Walter Zimmerman, "Main Results of the 'Telome Theory,'" *Paleobotanist* 1 (1952): 456–470.
12. Andrews, journal, 20–21 July 1968.
13. Kasper, "Pertica," 898.
14. Ibid., 897.
15. William H. Forbes to Arthur J. Boucot, 24 January 1970, Smithsonian Institution Archives. Accession 15-182, Box 2, Folder 49: Forbes, William H., 1961-1970.

16. Arthur J. Boucot to Arthur Mraz, 29 January 1970, Smithsonian Institution Archives. Accession 15-182, Box 2, Folder 49: Forbes, William H., 1961-1970.
17. Gary M. Boone to Dean of Faculties, Aroostook State College, 21 January 1970, Smithsonian Institution Archives. Accession 15-182, Box 2, Folder 49: Forbes, William H., 1961-1970.
18. Bradford A. Hall to "to whom it may concern," 22 January 1970, Smithsonian Institution Archives. Accession 15-182, Box 2, Folder 49: Forbes, William H., 1961-1970.

9. The Maine State Fossil

1. Clark T. Irwin, Jr., telephone interview with author, 21 September 1987.
2. Clark T. Irwin, Jr., "State lags behind fossil gap," *Portland Press Herald*, 22 June 1984.
3. Ibid.
4. Ibid.
5. Clark T. Irwin, Jr., "Fossil expert Forbes as rare as his subject," *Portland Press Herald*, 22 June 1984.
6. Irwin, Jr., "State lags."
7. William H. Forbes, interview with author, 17 September 1987.
8. William H. Forbes, *Significant Bedrock Fossil Localities in Maine and Their Relevance to the Critical Areas Program*, Planning Report 46 (Augusta: Maine Critical Areas Program, State Planning Office, 1977), 37.
9. Ibid., 1.
10. Ibid., 70–74.
11. Legislative Record of the One Hundred and Twelfth Legislature of the State of Maine, Volume 1, First Regular Session, December 5, 1984–June 20, 1985, House, 29 January 1985 (Augusta, Maine: Maine State Law and Legislative Reference Library), 101, http://legislature.maine.gov/lawlib.
12. Ibid., Senate, 30 January 1985, 110.
13. Information on members on the Committee on State Government came from David Rawson, "112 Legislature will consider more than 2,000 bills Thursday," *Bangor Daily* News, 20 December 1984.
14. Michael Mokrzyki, "Maine coon cat proposed as state feline," *Lewiston Daily Sun*, 15 February 1985.
15. Ibid.
16. Walter A. Anderson, "Testimony Supporting LD 256 An Act to Designate 'Pertica Quadrifaria' the official fossil of the State of Maine," before the Joint Standing Committee on State Government, 14 February 1985. In possession of the authors.
17. Egon Schartel to Dan A. Gwadosky, 20 February 1985. In possession of the authors.
18. Maine Legislative Record, 112th, House, 28 February 1985, 202.
19. The Associated Press, "Fossil bill signed," *Kennebec Journal*, 14 March 1985.
20. Maine Legislative Record, 112th, House, 22 March 1985, 311.

21. Carolyn A. Lepage, "*Pertica quadrifaria*, Maine's State Fossil," Augusta, Maine: Maine Geological Survey, Department of Conservation, 1985.

10. New Revelations

1. Andrew E. Kasper, Jr., Patricia G. Gensel, William H. Forbes, and Henry N. Andrews, Jr., "Plant Paleontology in the State of Maine—A Review," *Maine Geological Survey, Studies in Maine Geology*, Vol. I (Augusta, Maine: Maine Geological Survey, 1988), 109–128.
2. Ibid., 109–110, 113–115, 127.
3. Ibid., 110–111, 114–120; see also Lisa Churchill-Dickson, *Maine's Fossil Record: The Paleozoic* (Augusta, Maine: Maine Department of Conservation, Maine Geological Survey, 2007), 280.
4. Kasper, et al., *Review*, 127.
5. See Robert Angelo Gastaldo, "Curriculum Vitae," Colby College, Maine. www.personal.Colby.edu/~ragastal/Gastaldo_Colby17_18_CV.pdf
6. Robert A. Gastaldo, interview with the authors, 16 September 2019.
7. Ibid.
8. Ibid.
9. See Robert E. Nelson, "Curriculum Vitae," Colby College, Maine. www.personal.colby.edu/geology/gifs/CV_REN_2014.pdf
10. See National Science Foundation, Award Abstract #0087433, RUI: Analysis of an Early-Middle Devonian Ecosystem: Trout Valley Formation, Maine. https://www.nsf.gov/awardsearch/showAward?AWD_ID=0087433
11. Jonathan P. Allen, et al., "Sedimentalogical Framework of the Trout Valley Formation, Middle Devonian, Maine," Geological Society of America, Annual Meeting, November 5–8, 2001. Paper No. 30-0. https://gsa.cofex.com/gsa/2001AM/finalprogram/abstract_23516.htm
12. Robert A. Gastaldo, et al., "Early Middle Devonian Terrestrial Ecosystems of Maine: Trout Valley Formation Revisited," Geological Society of America Annual Meeting, March 12-14, 2001, Paper No. 27-0. https://gsa.confex.com/gsa/2001NE/finalprogram/abstract_2603.htm
13. Robert E. Nelson, et al., "Early Middle Devonian Arthropod Remains from the Trout Valley Formation of North Central Maine, USA," Geological Society of America Annual Meeting, November 5-8, 2001, Paper No. 112-0. https://gsa.confex.com/gsa/2001AM/finalprogram/abstract_21270.htm
14. M. G. Terkla, J. P. Allen, R.E. Nelson, and R.A. Gastaldo, "Lower Middle Devonian Eurypterid Remains from the Trout Valley Formation of North Central Maine," *Geological Society of Maine Newsletter* 28, no. 2 (2002): 8.
15. Robert W. Selover, Robert A. Gastaldo, and Robert E. Nelson. "An Estuarine Assemblage from the Middle Devonian Trout Valley Formation of Northern Maine," *Palaios* (20) no. 2 (April 2005): 192–193.
16. Ibid., 192–196.
17. Jonathan P. Allen and Robert A. Gastaldo, "Sedimentology and Taphonomy of the Early to Middle Devonian Plant-bearing Beds of the Trout Valley Formation, Maine,"

in Greb, S. F. and DiMichele, W. A., *Wetlands Through Time: Geological Society of America Special Paper* (Boulder, Colorado: Geological Society of America, 2006), 58.
18. Ibid., 58–59.
19. Ibid., 59–61, 73–75.
20. Ibid., 72, 74–75.
21. Kasper, et al., *Review*, 115.
22. Robert A. Gastaldo, "New Paleontological Insights into the Emsian-Eifelian Trout Valley Formation, Baxter State Park's Scientific Forest Management Area, Aroostook County, Maine," *Palaios* v. 3 (2016):339–340.
23. Ibid., 340.
24. Ibid., 340–341, 345.
25. Ibid., 346.
26. Gastaldo, interview, 16 September 2019.

11. The Descendent

1. David Domozuch, Iben Serensen, and Zoe A. Popper, "(Editorial) Charophytes: Evolutionary Ancestors of Plants and Emerging Models for Plant Research," *Frontiers in Plant Science* 8 (March 2017):338.
2. James D. Mauseth, *Botany: An Introduction to Plant Biology*, Fifth Edition (Burlington, Mass.: Jones & Bartlett Learning, 2014), 513, 529–530; Andrew E. Kasper, Jr., "A New Genus of Devonian Fossil Plants from Northern Maine," Dissertation for the degree of Doctor of Philosphy (Storrs, Connecticut, University of Connecticut, 1970); and Lisa Churchill-Dickson, *Maine's Fossil Record: The Paleozoic* (Augusta, Maine: Maine Department of Conservation, Maine Geological Survey, 2007), 280–82; and Andrew E. Kasper to the authors, 1 April 2020.
3. Wilson N. Stewart and Gar W. Rothwell, *Paleobotany and the Evolution of Plants* (Cambridge, England: Cambridge University Press, 1993), 275.
4. Abstract Publisher Summary for Chapter 12, "Progymnosperms," in Thomas N. Taylor, Edith L. Taylor, and Michael Krings, *Paleobotany: The Biology and Evolution of Fossil Plants*, 2nd edition (Academic Press: Cambridge, Massachusetts, 2009). See also sciencedirect.com/book/9780123739728/paleobotany#book-info.
5. Henry N. Andrews, Patricia G. Gensel, and Andrew E. Kasper, "A New Fossil Plant of Probable Intermediate Affinities (Trimerophyte-Progymnosperm)," *Canadian Journal of Botany* 53 (1975):1719–1728; and Charles H. Wellman and Patricia G. Gensel, "Morphology and Wall Ultrastructure of the Spores of the Lower Devonian Plant *Oocampsa catheta* Andrews et al., 1975," *Review of Paleobotany and Palynology* 130 (2007):269–295.
6. See Mauseth, *Botany*, 530; "Introduction to the Progymnosperms," www.ucmp.edu/seedplants/progymnosperms.html; Brigitte Meyer-Berthaud, Stephen Scheckler, and Jobst Wendt, "Archaeopteris is the Earliest Known Modern Tree," *Nature* 398 (no. 6729) (April 1999) 700–701; and "Progymnosperms," *A Dictionary of Biology* (Encylopedia.com, 17 January 2019). https//www.encyclopedia.com/science/dictionaries-thesaures-pictures-and-pressreleases/progymnosperms-0.

7. C. M. Govil, *Gymnosperms: Extinct and Extant* (Delhi, India: Krishna Prakashan Media, 2007), 148.
8. See "Cordaitales," Wikipedia. https://en.wikipedia.org/index.php?title=Cordaites&oldid=842479021
9. Mauseth, *Botany*, 533–535.
10. Fay Hyland, *The Conifers of Maine,* bul. 345 (Orono, Maine: Cooperative Extension Service, University of Maine, n.d.), 2; and *Forest Trees of Maine* bul. 24, Tenth Edition (Augusta, Maine: Maine Forestry Department, 1973), 2–7.
11. Dean B. Bennett, *The Forgotten Nature of New England: A Search for Traces of the Original Wilderness* (Camden, Maine: Down East Books, 1996), 286.
12. "Evangeline: A Tale of Acadie," Henry Wadsworth Longfellow. https://nslegislature.ca/sites/default/files/pdfs/about/evangeline/Evangeline.pdf
13. Henry David Thoreau, "Walking," *The Atlantic Monthly* 9 (June 1862): 657–674.
14. Richard G. Wood, *A History of Lumbering in Maine, 1820-1861* (Orono, Maine: University of Maine Press, 1972), 19.
15. William Cronon, *Nature's Metropolis: Chicago and the Great West* (New York, NY: W. W. Norton and Company, 1991), 183.
16. Environmental Studies, Lake Forest College. https//www.lakeforest.edu/academics/programs/environmental/. . ./pinus_strobus.php
17. Governor Percival Baxter, "Baxter Communications (1955) to Governor Muskie, Senate and House of Representatives, 97th Legislature." https://baxterstatepark.org/wp-content. . ./SFMAForestManagementPlan1988.pdf
18. John W. Hakola, *Legacy of a Lifetime: The Story of Baxter State Park* (Woolwich, Maine: TBW Books, 1981), 305–306.
19. John Herrick, "Percival Baxter's Scientific Forestry Management Area," Forest Ecology Network, *The Maine Woods* (Fall 1999). www.forestecologynetwork.org/tmwfall1999-11.htm/

12. A Perspective on Science and Nature

1. Natural Areas Program, "Baxter Park Region Focus Area of Statewide Ecological Significance" (Augusta, Maine: Natural Areas Program, Maine Department of Agriculture, Conservation, and Forestry, n.d.).
2. See Baxter State Park paper, "Climate Change in Baxter State Park." https://www.baxterstatepark.org/wp-content/uploads/2017/05/Park-CC-overview_4_1_15.pdf
3. International Union for the Conservation of Nature (IUCN), *Issues Brief: Forests and Climate Change.* http://www.iucn.org/resources/issues/issues-briefs-forests-and-climate-change
4. See "Prehistoric Climate Change and Why It Matters Today," Smithsonian Institution, Smithsonian Center for Education and Museum Studies. http://www.smithsonianeducation.org/idealabs/prehistoric_climate_change/index.htm
5. "Ancient Fossils and Modern Climate Change: The Work of Jennifer McElwain," Understanding Evolution. http://evolution.berkeley.edu/evolutionary/article/mcelwain_01

6. Margret Steinthorsdottir, et al., "Fossil Plant Stomata Indicate Decreasing Atmospheric CO_2 Prior to the Eocene-Oligocene Boundary," *Climate of the Past* 12 (2016): 439–454.
7. Linda Lear, "The Next Page: Rachel Carson, A Child of the Allegheny," *Pittsburgh Post-Gazette*, 7 July 2018.
8. Linda Engelsiepen, "Rachel Carson: Giving Nature a Voice," *Ecology Webinar Series* (26 May2010). www.ecology.com/2010/o5/28/rachel-carson?
9. Tom L. Phillips, "Henry Nathaniel Andrews, Jr., 1910-2002," in *Biographical Memoirs* 88 (Washington, D.C.: National Academy of Sciences, 2006), 15.
10. William H. Forbes, *Significant Bedrock Localities in Maine and Their Relevance to the Critical Areas Program*, Planning Report 46 (Augusta, Maine: Maine Critical Areas Program, State Planning Office, 1977), 91–94.

EPILOGUE

1. Glen H. Mittelhauser, et al., *The Plants of Baxter State Park* (Orono, Maine: University of Maine Press, 2016).

INDEX

ABOUT THE AUTHORS

Dean B. Bennett, Professor Emeritus at the University of Maine at Farmington, writes and illustrates books about nature, wilderness, and human relationships with the natural world. His book *Ghost Buck: The Legacy of One Man's Family and Its Hunting Traditions*, published by Islandport Press in 2015, won the John Cole Award, a Maine Literary Award. His other books include *Maine's Natural Heritage: Rare Species and Unique Natural Features*, 1988, Down East Books; *Allagash: Maine's Wild and Scenic River*, 1994, Down East Books; *The Forgotten Nature of New England: A Search for Traces of the Original Wilderness*, 1996, Down East Books; *Nature and Renewal: Wild River Valley and Beyond*, 2009, Tilbury House Publishers. His book *The Wilderness from Chamberlain Farm: A Story of Hope for the American Wild*, Island Press, was awarded first place in the environment category for 2001 books by *ForeWord Magazine*. He also wrote and illustrated three children's books, published by Down East Books: *Everybody Needs a Hideaway*, 2004; *Finding a Friend in the Forest*, 2005; and *The Late Loon*, 2006. Two received recognition by the *Chicago Tribune* and the National Wildlife Federation. His professional writing includes a book for the United Nations Educational, Scientific and Cultural Organization on the evaluation of environmental learning and an invited chapter in the international organization's first book on environmental education.

Bennett was born and raised in western Maine. His writing draws on a diverse background of experience and education related to people and their environment. He holds a journeyman's certificate in cabinetmaking and architectural millwork from the Maine State Apprenticeship Council, a BS in Industrial Arts Education, an MS in Science Education, and from the University of Michigan, a PhD in Resource Planning and Conservation. His career spans teaching all grade levels in public schools, work in state government, and many years as professor of science education at the University of Maine at Farmington. For nearly a decade he taught natural history and environmental education summer courses for teachers at Maine's Conservation School in Bryant Pond.

Bennett has received several awards for his scholarship, teaching, and conservation work. He enjoys canoeing, hiking, deer hunting, playing a four-string jazz banjo, and playing in a steel drum band at the University of Maine at Augusta. He lives in Hallowell, Maine, with his wife, Sheila.

Sheila K. Bennett is Professor Emeritus at the University of Maine at Augusta where she taught in the science department for 36 years. She pioneered a laboratory science course at a distance, Introduction to the Natural Sciences, using instructional television to broadcast to students located statewide. Before retirement she developed a laboratory science course available to online students worldwide.

She was the recipient of the Libra Professorship and Distinguished Scholar's award which allowed her to study ancient environments of Maine through fossil evidence. She incorporated what she learned in the courses she taught. She and her husband, Dean, were selected to give the Brandwein Lecture at the 2014 National Science Teachers Association conference in Atlanta, Georgia. Their topic was: Regarding the Ecology of Science Education: Connections to Environmental and Distance Education.

She coauthored with Dean Bennett *Allagash Wilderness Waterway: A Natural History Guide* published by the Bureau of Parks and Recreation, Maine Department of Conservation—a pocket guide for visitors to the Waterway. In 2014 she and her husband received the Maine Conservation Voters' Harrison Richardson Environmental Leadership award which recognized their dedication to *teaching, writing, speaking, advocating, and inspiring all of us to care for the nature of Maine and wild places.*

She holds a baccalaureate degree in medical technology from the University of Vermont, a master's in science education, and a PhD in biological sciences from the University of Maine. She worked in hospitals in Ithaca, New York and in Denver, Colorado. She was acting director of the Guam Public Health Laboratory, Agana, Guam.

In 1972 she was appointed by Governor Kenneth M. Curtis to serve on the Task Force on Energy, Heavy Industry, and the Maine Coast. She served as President of the Maine Association of Conservation Commissions. She taught science in middle school for the Gardiner Public School System and Environmental Education at the elementary level in the Waterville Public School System.

She lives in Hallowell, Maine, with her husband, Dean, and enjoys canoeing, walking in the woods, and playing in the University of Maine at Augusta community steel-drum band ensemble, Vintage Steel.

CPSIA information can be obtained
at www.ICGtesting.com
Printed in the USA
BVHW041156260622
640104BV00009B/2